THE TRANSIENT MILKY WAY: A PERSPECTIVE FOR MIRAX

To learn more about the AIP Conference Proceedings, including the Conference Proceedings Series, please visit the webpage
http://proceedings.aip.org/proceedings

THE TRANSIENT MILKY WAY: A PERSPECTIVE FOR MIRAX

São José dos Campos, Brazil   7 - 9 December 2005

EDITORS
Flavio D'Amico
João Braga
National Institute for Space Research (INPE)
São José dos Campos, Brazil

Richard E. Rothschild
University of California
San Diego, California, U.S.A.

SPONSORING ORGANIZATIONS
National Institute for Space Research (INPE)
Brazilian Space Agency (AEB)

Melville, New York, 2006
AIP CONFERENCE PROCEEDINGS ■ 840

Editors

Flavio D'Amico
João Braga
National Institute for Space Research (INPE)
Astrophysics Division
Avenida dos Astronautas 1758
12227-010 São José dos Campos
Brazil

E-mail: damico@das.inpe.br
  braga@das.inpe.br

Richard E. Rothschild
University of California, San Diego (UCSD)
Center for Astrophysics and Space Sciences (CASS)
9500 Gilman Drive
La Jolla, CA 92093-0424
USA

E-mail: rrothschild@uscd.edu

The article on pp. 45 - 49 was authored by a U.S. Government employee and is not covered by the below mentioned copyright.

Authorization to photocopy items for internal or personal use, beyond the free copying permitted under the 1978 U.S. Copyright Law (see statement below), is granted by the American Institute of Physics for users registered with the Copyright Clearance Center (CCC) Transactional Reporting Service, provided that the base fee of $23.00 per copy is paid directly to CCC, 222 Rosewood Drive, Danvers, MA 01923, USA. For those organizations that have been granted a photocopy license by CCC, a separate system of payment has been arranged. The fee code for users of the Transactional Reporting Services is: 0-7354-0332-5/06/$23.00

© 2006 American Institute of Physics

Permission is granted to quote from the AIP Conference Proceedings with the customary acknowledgment of the source. Republication of an article or portions thereof (e.g., extensive excerpts, figures, tables, etc.) in original form or in translation, as well as other types of reuse (e.g., in course packs) require formal permission from AIP and may be subject to fees. As a courtesy, the author of the original proceedings article should be informed of any request for republication/reuse. Permission may be obtained online using Rightslink. Locate the article online at http://proceedings.aip.org, then simply click on the Rightslink icon/"Permission for Reuse" link found in the article abstract. You may also address requests to: AIP Office of Rights and Permissions, Suite 1NO1, 2 Huntington Quadrangle, Melville, NY 11747-4502, USA; Fax:

L.C. Catalog Card No. 2006926495

ISBN 0-7354-0332-5
ISSN 0094-243X
Printed in the United States of America

CONTENTS

Preface vii

MIRAX MISSION

MIRAX Mission Overview and Status. 3
    J. Braga and the MIRAX Team
A Soft X-ray Imager for MIRAX 8
    J. in 't Zand, W. Mels, and J. Heise
Scientific Satellites at INPE 18
    M. A. Chamon and T. R. de Carvalho

MIRAX SCIENCE—X-RAY

IGR Sources and MIRAX 25
    J. A. Tomsick
The INTEGRAL Galactic Bulge Monitoring Program 30
    E. Kuulkers, S. E. Shaw, S. Brandt, J. Chenevez, T. J.-L. Courvoisier,
    K. Ebisawa, P. Kretschmar, C. B. Markwardt, N. Mowlavi, T. Oosterbroek,
    A. Orr, A. Paizis, C. Sanchez-Fernandez, and R. Wijnands
An INTEGRAL View of the Inner Galaxy 35
    R. Walter
Monitoring Neutron-star High Mass X-ray Binaries in the
INTEGRAL Galactic Plane Survey 40
    J. Wilms
RXTE Observations of Galactic Center Transients 45
    C. B. Markwardt
Accretion-powered Millisecond Pulsar Outbursts 50
    D. K. Galloway
What Can We Learn from Long-term Monitoring of X-ray Bursters? 55
    A. Cumming
High Mass X-ray Binaries Pulsars—A Brief Review at Hard X-rays 60
    A. Santangelo
Long-term Developments in Her X-1: Correlation Between the
Histories of the 35-day Turn-on Cycle and the 1.24 Sec Pulse Period 65
    R. Staubert, S. Schandl, D. Klochkov, J. Wilms, K. Postnov, and
    N. Shakura
Swift X-ray Telescope Observations of Galactic Transients 71
    J. A. Kennea
Can Black Holes Provide the Emitted Energy of GRBs? 76
    R. Opher

MIRAX SCIENCE—MULTIWAVELENGTH

Jets from Galactic X-ray Transients: the MIRAX Perspective 83
    E. Gallo and R. Fender

Multiwavelength Variability in Transient Black Hole Binaries 88
    R. I. Hynes

Infrared Counterparts to X-ray Sources in the Galactic Center Region 93
    F. Jablonski and L. A. Ramos

Search for an Infrared Counterpart of IGR J16358-4756 97
    F. D'Amico, F. Jablonski, C. V. Rodrigues, D. Cieslinski, and G. Hickel

MAGPIS: The Multi-array Galactic Plane Imaging Survey 102
    R. H. Becker, R. L. White, and D. J. Helfand

MIRAX INSTRUMENTS AND SOFTWARE

CZT Detector and HXI Development at CASS/UCSD 107
    R. E. Rothschild, J. A. Tomsick, J. L. Matteson, M. R. Pelling, and S. Suchy

HXI Imaging Simulations and Sensitivity 112
    J. Mejia and J. Braga

Event Pre-processor for the CZT Detector on MIRAX 117
    E. Kendziorra, T. Schanz, S. Suchy, and G. Distratis

MIRAX Software Aspects 122
    J. Wilms, S. Schwarzburg, R. Remillard, E. Kendziorra, R. Staubert, and R. E. Rothschild

On-board Computing Subsystem for MIRAX: Architectural and Interface Aspects 127
    V. Santiago

Payload Software Validation and Integration 132
    M. de Fátima Mattiello-Francisco

Workshop Program 137

List of Participants 141

Author Index 143

PREFACE

Nearly 60 scientists and engineers met in São José dos Campos, Brazil, for the workshop *"The Transient Milky Way: A Perspective for MIRAX"* in December 2005. The main goal of the meeting was the discussion of the science to be investigated with the launch of the MIRAX (Monitor e Imageador de RAios-X) X-ray astronomy satellite mission, scheduled for 2011. Along with the science presentations and discussions, some technical talks concerning the scientific instruments, satellite systems and mission software were presented as well. In assembing these proceedings, we have reordered the sequence of papers from that of the workshop in order to follow a more natural thematic sequence. We hope the readers will find this sequence more useful.

MIRAX is the first Brazilian project of an astronomical satellite and has very strong international partnership and collaboration, especially with the University of California San Diego, the Netherlands Institute for Space Research, the University of Tübingen in Germany and the Massachusetts Institute of Technology. The satellite platform will be built by INPE (National Institute for Space Research in Brazil) and the launch will be provided by the Brazilian Space Agency.

The science objectives of MIRAX include the detailed study of transient X-ray sources, hard X-ray surveys and explosive phenomena in astrophysics. With its broadband, wide field-of-view and quasi-continuous imaging spectroscopy observing approach, MIRAX will be able to uniquely address key points about the physics of accretion onto neutron stars and black holes, including X-ray burst and superburst recurrence times, detailed evolution of state transitions on accreting compact objects, accretion torque history on neutron stars, relativistic jets on microquasars and other systems, flaring X-ray sources and fast transients. The mission will also be capable of making important contributions to the study of gamma-ray bursts and X-ray flashes.

The observing strategy of MIRAX includes staring at a large fraction of the central Galactic plane for about 9 months per year. This will allow the detection, localization, identification and study of a large number of short-lived, unpredictable phenomena that are otherwise missed by low duty cycle observations such as the ones performed by Integral and Swift. Furthermore, MIRAX will be uniquely capable of studying longer-lived phenomena (in a timescale of months) in exquisite detail. Since MIRAX will be monitoring the central Galactic plane for extended periods of time, it will perform an important alert service for observations of transients in other wavelength bands. Simultaneous and/or follow-up optical and radio observations of MIRAX sources will be crucial for the detailed modelling of the emitting objects.

The Scientific Organizing Committee for the workshop consisted of J. Braga (INPE, Chair), F. D'Amico (INPE), J. Mejía (INPE), F. Jablonski (INPE), M. A. Chamon (INPE), R. Rothschild (UCSD), J. Heise (SRON), J. in 't Zand (SRON), R. Remillard (MIT), R. Staubert (IAAT), E. Kendziorra (IAAT), E. Kuulkers (ESAC), J. Wilms (U. Warwick) and E. Janot Pacheco (USP, Brazil). We gratefully acknowledge the assistance of Mônica de Oliveira and Alexandra Pinto during the meeting. We also thank Beatriz

Kozilek and Carlos Vieira for the design of the workshop poster.

Finally, a high point of the meeting was the happy hour that happened on Thursday night. The foreigners enjoyed the famous Brazilian drink "caipirinha" and it was a good opportunity for old and new friends to get together and relax.

João Braga
Conference Chair
March 2006

MIRAX MISSION

MIRAX Mission Overview and Status

João Braga and the MIRAX team

National Institute for Space Research, CP 515, S. J. Campos, SP Brazil 12201-970

Abstract. We present an overview of the planned MIRAX X-ray astronomy satellite mission. The main scientific mission drivers are discussed, and the proposed instruments and baseline parameters are described. Finally, we briefly describe the satellite platform and discuss the current status of the mission.

Keywords: X rays, satellite, coded aperture, transient X-ray sources
PACS: 95.40.+s, 95.55.-n, 95.55.Ka, 95.85.Nv

MISSION OVERVIEW

The Monitor e Imageador de Raios-X (MIRAX) is a small (~250 kg, ~240 W) X-ray astronomy satellite mission designed to monitor a large region around the central Galactic plane for transient phenomena. With a field-of-view of more than 1000 square degrees and an angular resolution of ~6 arcmin, MIRAX will provide an unprecedented discovery-space spectral coverage to study X-ray variability in detail, from fast X-ray novae to long-term (several months) variable phenomena. In comparison, observations of the Galactic Center region made by the Integral and Swift missions suffer from low duty cycles which make them unlikely to detect short-lived transients and unable to perform detailed studies of longer-lived phenomena. Chiefly among MIRAX science objectives is its capability of providing simultaneous complete temporal coverage of the evolution of a large number of accreting black holes, including a detailed characterization of the spectral state transitions in these systems. MIRAX's instruments will include a soft X-ray (2-18 keV) and two hard X-ray (10-200 keV) coded-aperture imagers, with sensitivities of ~5 and ~2.6 mCrab/day, respectively. The hard X-ray imagers (HXI) will be built at the Instituto Nacional de Pesquisas Espaciais (INPE), Brazil, in close collaboration with the Center for Astrophysics & Space Sciences (CASS) of the University of California, San Diego (UCSD) and the Institut für Astronomie und Astrophysik of the University of Tübingen (IAAT) in Germany; UCSD will provide the crossed strip position-sensitive (0.5-mm spatial resolution) CdZnTe (CZT) hard X-ray detectors. The soft X-ray imager (SXI), provided by the Netherland Institute for Space Research (SRON), will be the spare flight unit of the Wide Field Cameras that flew on the Italian-Dutch BeppoSAX satellite with excelent performance.

The MIRAX spacecraft will employ a 3-axis attitude control system. The three X-ray cameras will be co-aligned and inertially pointed to the central Galactic Plane for

9 months per year. Every object in the FOV will be observed for 60 min of every 90-min orbit, 15 times a day. During the other 3 months, the satellite will maneuver so the cameras will point to other selected targets. There will be no moving parts in the payload. The satellite platform will have an attitude precision of 0.5° with a 36"/hr stability (jitter) and 20" attitude knowledge given by an APS star camera.

MIRAX is an approved mission of the Brazilian Space Agency (Agência Espacial Brasileira - AEB) and is part of the Brazilian National Plan for Space Activities for 2005-2014. The satellite is scheduled to be launched in 2011 in a low-altitude (~550 km) circular equatorial orbit by the Brazilian launcher VLS. The mission will use a S-band telemetry with a expected rate of 1.5 Mbps, which will be enough to dump all the science data with no need for on-board compression. The ground station will be provided by INPE and located at Alcântara, Brazil (-2° latitude).

Table 1 shows the MIRAX baseline parameters.

TABLE 1. MIRAX mission baseline parameters

Mission and spacecraft parameters		
Mass	~250 kg (total), ~100 kg (payload)	
Power	~240 W (total), ~90 W (payload)	
Orbit	equatorial, circular, ~550 km	
Telemetry	S-band (2200-2290 MHz), ~1.5 Mbps downlink	
Launch	2011 by Brazilian VLS ("Veículo Lançador de Satélites")	
Instrument parameters	**Hard X-ray Imager (CXD)**	**Soft X-ray Imager (CXM)**
energy range	10-200 keV	2-28 keV
angular resolution	7.5 arcmin	5 arcmin
Localization	< 1 arcmin (10σ source)	< 1 arcmin (10σ source)
field-of-view	58° x 26° FWHM along the GP	20° x 20° FWHM
spectral resolution	< 5 keV @ 60 keV	1.2 keV @ 6 keV
time resolution	< 10 μs	122 μs
Sensitivity	< 2.6 mCrab (1 day, 5 σ)	< 5 mCrab (1 day, 5 σ)
Detector area	2 x 340 cm^2	520 cm^2

MIRAX INSTRUMENTS

The Hard X-Ray Imagers

The Hard X-Ray Imagers (HXIs) will be built in close collaboration with the Center for Astrophysics and Space Science (CASS) of UCSD and the Institute for Astronomy and Astrophysics (IAAT) of the University of Tuebingen in Germany. The HXIs will operate from 10 to 200 keV. The detector plane of each camera will be a 3 x 3 array of state-of-the-art CdZnTe crossed-strip detector modules with 0.5 mm spatial resolution developed at CASS, with a total area of 340 cm^2. Each detector module is a 2 x 2 array of 32 mm x 32 mm x 2mm thick CZT detectors. The detectors will be surrounded by an active plastic scintillator shield and by a passive

Pb-Sn-Cu graded shield. A 315 mm x 275 mm Tungsten coded-mask with 1.3 mm-side square cells (0.5 mm-thick) will be placed 600 mm away from the detector to provide images with 7'30" angular resolution. The basic pattern of the mask will be a 139 x 139 Modified Uniformly Redundant Array (MURA – [1] [2]) which will allow for full shadowgrams to be cast on the position-sensitive detector area and will provide no source ambiguity problems in the fully-coded field-of-view (FCFOV). A sketch of the HXI is shown in Fig. 1.

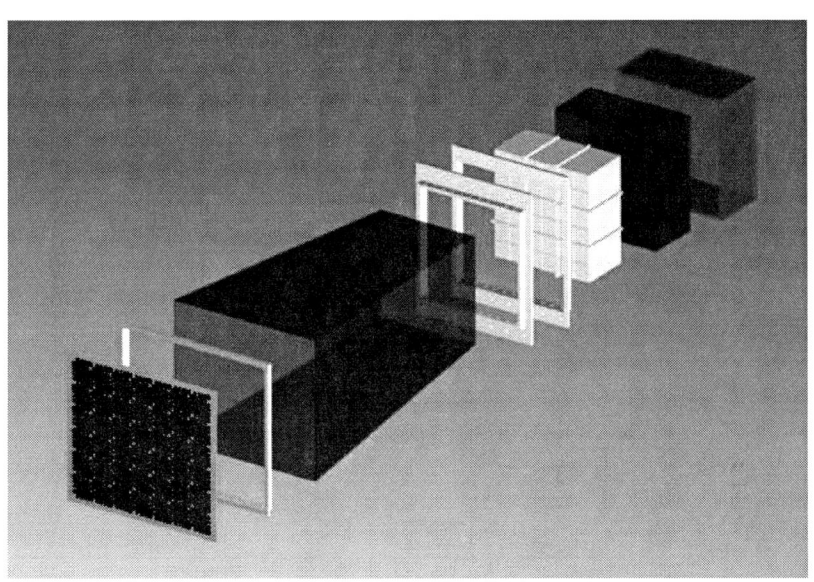

FIGURE 1. Exploded diagram of the MIRAX hard X-ray camera. From left to right, the elements are: coded-mask, coded-mask support structure, Pb-Sn-Cu passive-shield walls, two structural flanges, detector modules, Pb-Sn-Cu passive shield and plastic scintillator active shield.

The pointing axes of the two HXIs will be offset by an angle of 29° in order to provide a uniform sensitivity over a 39° FCFOV in one direction; the perpendicular direction will have a 6°12' FCFOV. In such a configuration the FWHM FOV is 58° x 26°. During the observations of central Galactic Plane, the wider direction of the FOV will be aligned with the GP. Figure 2 shows the fractional coded-area (considering the two cameras) as a function of angle, for the direction aligned with the GP.

The Soft X-Ray Imager

The SXI, provided by SRON, is the spare flight unit of the Wide Field Cameras (WFCs – [3]) of the *BeppoSAX* mission [4], and will operate from 1.8 to 28 keV.

The SXI will have a 5' angular resolution in a 20° x 20° FWHM FOV. The addition of the WFC to the MIRAX payload will provide soft X- ray spectral coverage which will be extremely important for the study of the several classes of sources in the MIRAX FOV. Furthermore, the excellent performance of the WFCs on *BeppoSAX* brings to MIRAX an instrument that has already been tested and used successfully in orbit with very little degradation on a time scale of several years.

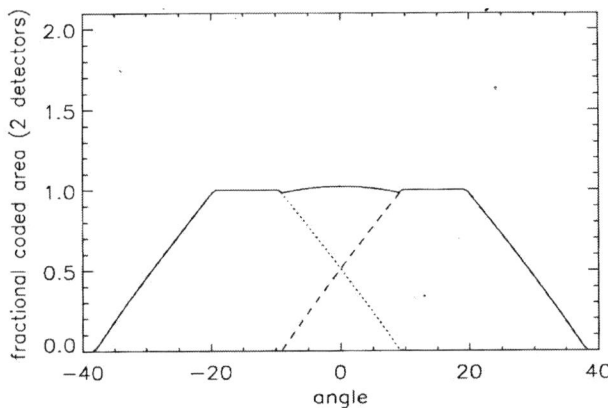

FIGURE 2. The fractional coded area of the two hard X-ray cameras of MIRAX with the main axes offset by 29°. This angle provides a nearly uniform FCFOV of 39° along the GP.

A preliminary design of the MIRAX spacecraft is show on Fig. 3. The SXI is mounted on top of the two HXIs, which are offset by 29°. A star camera with an Active Pixel Sensor (APS), currently being developed at INPE, is placed in the space between the two CXDs.

SPACECRAFT AND MISSION OPERATIONS

The MIRAX spacecraft will be based on the satellite bus being developed at INPE. The platform employs a 3-axis attitude stabilization system with 2 star trackers, a sun sensor and a magnetometer. Torque rods and reaction wheels will be used as attitude controllers. The MIRAX payload will have no moving parts and a mass of ~100 kg, while the total spacecraft mass is expected to be under 250 kg. There will be no propulsion and the pointing will be inertial. The pointing precision will be 0.5°, with 36"/hr stability (1/10 of the image pixel) and 20" attitude knowledge. The power consumption of the payload will be between 88 and 96 W, depending on the final configuration and the electronics requirements, and the total power of the satellite will be around 240 W.

The MIRAX mission duration is required to be 2 years, with a possible extension to 5 years. A ground station at Natal, Brazil, operated by INPE, will be assembled. Possibly, a second ground station in Kenya will be available. The space operation S-band (2200 – 2290 MHz) will be used for downlink and command uplink. It is expected that downlink data rates up to ~2 Mbits/s will be possible to reach, depending on the modulation and on coordination with other satellites. Our current estimates indicate that a rate of 1.5 Mbits/s will be enough to dump all the data with no compression if we use one station.

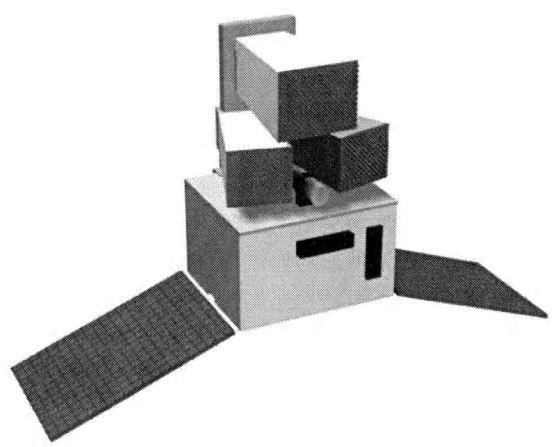

FIGURE 3. A preliminary view of the MIRAX spacecraft. The two cameras mounted over the satellite bus are the hard X-ray cameras (HXIs), whereas the soft X-ray camera is on top of the HXIs. The APS optical star camera is placed in between the HXIs. The external dimensions are approximately 1.5 m x 0.7 m x 0.7 m (with the solar panels folded over the spacecraft).

MIRAX is expected to be launched by 2011 by the Brazilian satellite launcher VLS, in case it is tested successfully and is officially considered a reliable launcher by the Brazilian Space Agency. In case a VLS is not available, other possibilities will be considered, such as a Pegasus launch or as a piggy-back payload on larger launchers.

MIRAX data will be 100% available to the community immediately. Databases will be setup at the missions centers in Brazil (INPE) and at UCSD. The database will also be available at HEASARC (Goddard Space Flight Center). Specific webpages with several data products will be available.

REFERENCES

1. S. R. Gottesman and E. E. Fenimore, *Applied Optics*, **28**, 4344-4352 (1989).
2. J. Braga, J., F. D'Amico, T. Villela, J. Mejía, R. A. Fonseca, and E. Rinke, *Rev. Sci. Instr.*, **73**, 3619-3628 (2002).
3. R. Jager, W. A. Mels, A. C. Brinkman, M. Y. Galama, H. Goulooze, J. Heise, P. Lowes, J. M. Muller, A. Naber, A. Rook, R. Schuurhof, J. J. Schuurmans, and G. Wiersma, *Astr. Astrophys. Suppl. Series,* **125**, 557-572 (1997).
4. Boella, G., R. C. Butler, G. C. Perola, et al., *Astr. Astrophys. Suppl. Series*, **122**, 299-307 (1997).

A Soft X-ray Imager for MIRAX

Jean in 't Zand, Wim Mels & John Heise

SRON Netherlands Institute for Space Research,
Sorbonnelaan 2, 3584 CA Utrecht, the Netherlands

Abstract. The flight spare model of the BeppoSAX Wide Field Cameras is being considered as the Soft X-ray Imager for MIRAX. A description is provided of this instrument, the performance of its siblings on BeppoSAX, and the prospects of flying it on MIRAX. Like on BeppoSAX, the instrument on MIRAX will excel in the study of transient phenomena lasting shorter than 1 day.

Keywords: X- and γ-ray telescopes and instrumentation; flare stars; γ-ray bursts; X-ray binaries
PACS: 95.55.Ka; 95.85.Nv; 97.30.Nr; 97.70.Rz; 97.80.Jp

INTRODUCTION

From 1983 to 1995 the SRON Netherlands Institute for Space Research designed and constructed two wide field X-ray cameras (WFCs) for the Italian-Dutch 'Satellite per Astronomia X' (SAX, later BeppoSAX), as well as a prototype mission (COMIS/TTM) on the *Mir* space station (for a historic perspective, see [1]). In total three space-qualified BeppoSAX cameras were built of which one was assigned 'flight spare model' (FSM), to act as a backup should a problem occur with one of the flight models. Such a problem never occurred and the spare model was saved. It was subsequently preserved in a conditioned environment at SRON. Several times, the camera underwent tests showing the absence of any degradation.

Since 2003 the spare model is being considered for flight on the *Monitor e Imageador de RAios-X* (MIRAX) mission [2]. MIRAX is an X-ray mission being developed by the high-energy astrophysics group of the Astrophysics Division (DAS) of the National Institute for Space Research (INPE) in Brazil, in collaboration with the University of California at San Diego, the University of Tübingen and SRON. The advantage of using the WFC flight spare model is the minor investment needed to fly a good instrument. The combination of proven robust technology and still up-to-date capabilities of the instrument, in comparison with accepted programs in the same time frame, provide a welcome opportunity for meaningful and exciting science.

In this paper we present the case for flying the WFC FSM on MIRAX. We describe the instrument and the performance of its siblings on BeppoSAX. MIRAX will for 75% target the Galactic bulge. Therefore, it is insightful to discuss the results of such observations on BeppoSAX. These observations were one of the two most important successes of BeppoSAX-WFC, the other being the quick localization of γ-ray bursts. The latter, as well as other fast X-ray transients, will also be briefly discussed. We close with the implications of BeppoSAX results for the prospects of MIRAX.

 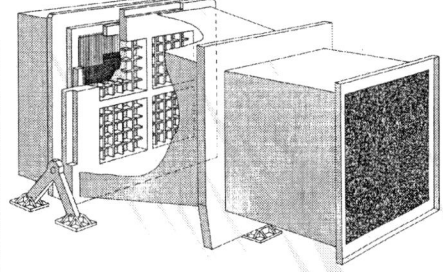

FIGURE 1. (Left) 1995 photograph of the 3 space-qualified WFCs. (Right) Schematic drawing of one WFC. The shielding and detector are cut away to reveal the support structure of the detector entrance window and the detector wire frames.

INSTRUMENT DESCRIPTION

general

The instrument has been described in detail elsewhere [3, 4]. We suffice with a few important points. A schematic drawing is provided in Fig. 1. The camera is a coded-aperture imager, with a mask and a detector with equal effective sizes of 255×255 mm^2, separated by 704 mm through a shielding connecting both. The shielding is opaque to X-ray photons so that only those celestial photons reach the detector that have gone through open parts of the mask. The mask consists of a rectangular pattern of closed and open elements with a grid spacing of 1 mm. 33% of the pattern is open [5]. The mask plate is made of gold-coated stainless steel in which the open elements are etched. To support isolated closed elements, a 0.1 mm wide support grid is left so that all open elements are in fact 0.9×0.9 mm^2 in size. The mask-detector configuration implies a field of view of $40° \times 40°$ (full-width at zero response) and an on-axis angular resolution of $5.'0$ (full width at half maximum; including the spatial resolution of the detector).

detector

The detector is a multi-wire position-sensitive proportional counter [3] filled at 2.2 bar with 95% Xe. The whole instrument weighs 42.5 kg. The complete titanium detector housing measures $40 \times 40 \times 5$ cm^3, while the 150 μm thick Be entrance window measures 255×255 mm^2. The entrance window is supported against the pressure difference by a Be structure up to 22 mm high (see Fig. 1). The combination of the photon mean free path in the gas and the transparency of the entrance window yields an effective bandpass of 2 to 30 keV. The 44 mm deep detector is divided in a main counter (upper 32 mm) and a guard counter (lower 12 mm) which is used for anti-coincidence against particles and high-energy photons scattered in the detector and satellite. The main counter has two cathode frames consisting of 50 μm thick wires 0.6 mm apart (at 900 V) and, in between, an anode frame of 20 μm thick wires with a pitch of 3.0 mm (at 3750 V). In total more than 1600 wires were manually installed. The cathodes are used for measuring the position of each photon, and the anode for measuring the energy. The cathode wires have an interconnection scheme delivering 35 signal outputs, multiplexing the position information. Apart from the guard counter, also the edges of the wire frames serve as anti-coincidence devices. The power consumption is 14.1 W. The event processor's speed can handle 2000 events per second at maximum (cf., the Crab source loads the detector with 330 phot s^{-1} when on-axis).

imaging concept

In coded aperture imaging each X-ray star casts a shadow of the mask on the detector. Since all stars are effectively at infinity, the mask shadows are all equal in size. Each shadow is merely characterized by its position (dictated by the star's off-axis position) and its strength (as a result of the star's brightness). Coded aperture cameras do not focus and photons from the star will end up over large parts of the detector, thus experiencing accumulated background levels. Furthermore, many stars in the field of view will use the same parts of the detector, introducing statistical cross talk. Therefore, coded aperture cameras are less sensitive than focusing telescopes with the same light collecting area. This disadvantage is compensated in X-rays by the ability to employ arbitrary large field of views (and by cost!).

The aperture/detector configuration employed in WFC is of the so-called 'box' type [6]: both camera elements are equal in size. This implies that there is no part of the field of view that is fully coded like in INTEGRAL-IBIS (e.g., [7]). The advantages of a box-type camera are that it is more compact for a given field of view, there is less statistical cross talk between the signals of different X-ray stars, and the background-induced noise will be smaller. The only disadvantage, lack of optimum imaging capability for extended sources, is considered less important because such sources do not belong to the primary objective of the BeppoSAX or MIRAX mission.

The nature of the detector has one important consequence for the imaging of point sources: due to the large off-axis angles allowed in this instrument, the photon penetration into the 32-mm deep main counter will leave streaks on the reconstructed image. The mean free path varies from less than 6 mm below 10 keV to 35 mm at 20 keV which, for an off-axis angle of 10°, will project exponential tails on the reconstructed sky of 5 and 30′ respectively. This clearly visible effect will have to be allowed for in extracting point source fluxes through modelling of the point spread function.

PERFORMANCE ON BEPPOSAX

Two WFCs were launched on BeppoSAX on April 30, 1996 and were in operation between June 4, 1996, and April 30, 2002. Every time the satellite traversed the South Atlantic Geomagnetic Anomaly the high-voltage of both cameras was switched off. This happened about 24,400 times. The net HV-on time is about 3.5 yr per camera. The gain evolution of the two cameras is plotted in Fig. 2. In 6 years of operation, the gain changed

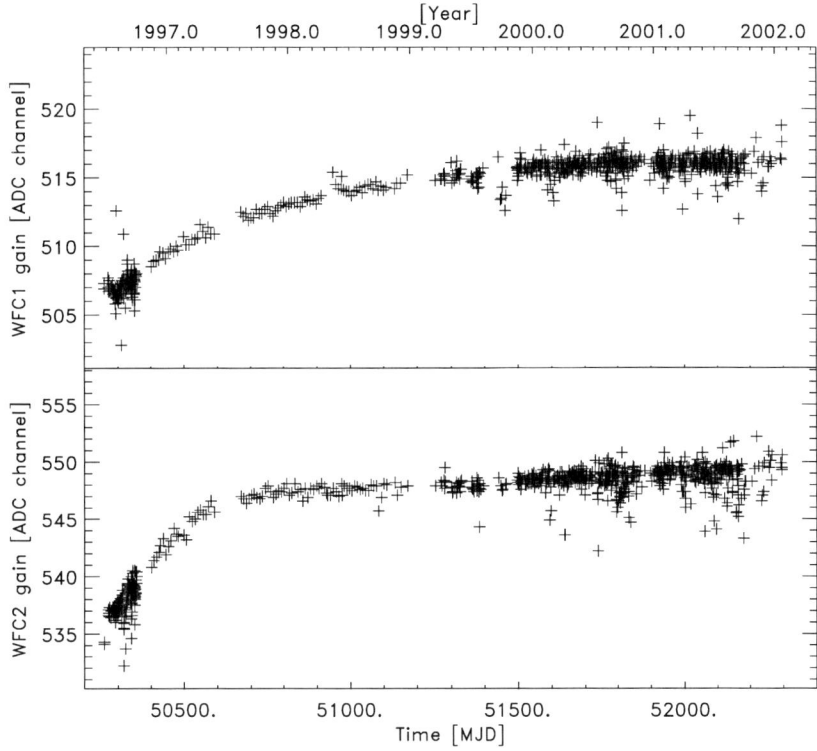

FIGURE 2. Gain trend as measured with Fe in-flight calibration sources.

by only 2% in both cameras. This is a good, stable and consistent performance between both cameras. This also applies to positional drift, spectral resolution and absolute response.

THE BEPPOSAX CAMPAIGN ON THE GALACTIC BULGE

The two BeppoSAX WFCs pointed in opposite directions and perpendicular to the Narrow Field Instruments. The observation program of the BeppoSAX WFCs was dictated by that of the Narrow Field Instruments and the sun angle of the solar panels. There was only one observation program primary to the WFCs: the campaign on the Galactic bulge. Combined with serendipitous pointings, a total of 6.2 Msec of data from both WFCs were acquired on the bulge (Fig. 3). This includes times when the field of view did not contain the earth and when the attitude control was optimum. If the latter requirement is relaxed, the exposure time increases by a few tens of percents. Such non-optimum data can still be of value for the study of short-duration X-ray flares. An

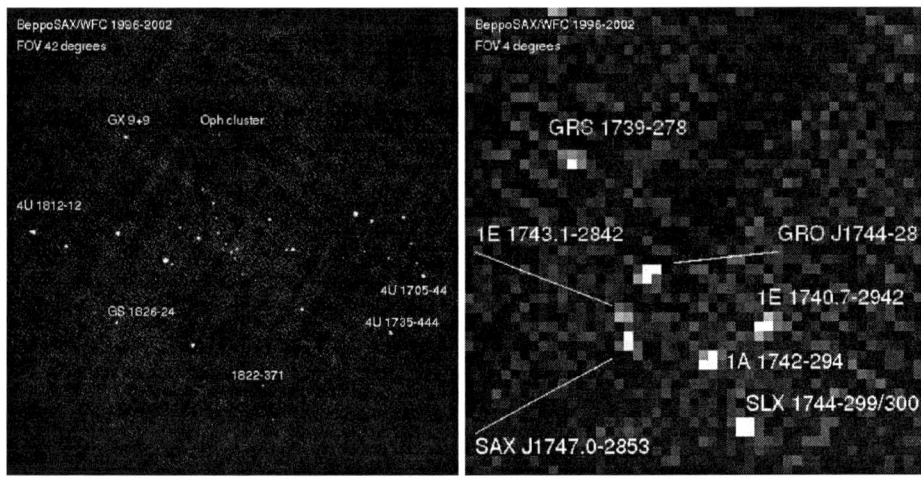

FIGURE 3. Complete field of view of WFC when targeted at the Galactic center (left) and zoomed (right). These images were generated by combining all 6.2 Msec worth of data.

overview of the Galactic bulge observations is given in [8]. Here we provide a summary.

transients

The BeppoSAX WFCs detected 34 X-ray transients in the Galactic Bulge between 1996 and 2002; only two were missed that were seen with other observatories (in particular RXTE). 21 transients were never seen before. Most transients are low-mass X-ray binaries. These outbursts are due to accretion disk instabilities due to the low mass-transfer rate from the donor star (e.g., [9]). Only SAX J1819.3-2525 and IGR J17544-2619 appear to be high-mass X-ray binaries. It is remarkable that only 8 out of the 36 transients reached peak fluxes in excess of 0.2 Crab units, suggesting peak luminosities that are considerably below the Eddington limit if one applies the canonical Galactic center distance of 8 kpc. Further interesting statistics from these X-ray transients are:

- The average transient rate is once every 2 to 3 weeks
- The measured recurrence times of transients LMXBs are above 0.4 yr.
- 80% of all LMXB transients have neutron star accretors and 20% black hole.
- 20% are 'long duration' transients, lasting longer than 1 yr (e.g., KS 1731-260, 2S 1711-339).
- 60% are fainter than 100 mCrab, 20% fainter than 20 mCrab (e.g., SAX J1806.5-2215).

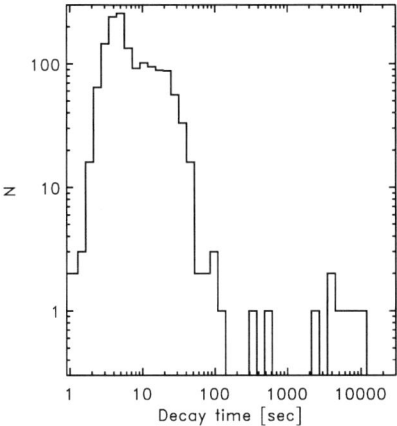

FIGURE 4. Histogram of e-folding decay times of those 1450 WFC-detected X-ray bursts that allow an accurate enough measurements.

X-ray bursts

A major contribution of the BeppoSAX-WFCs was in the field of type-I X-ray bursts. Such bursts are due to thermonuclear shell flashes in the upper layers of a neutron star. Thus, detection of an X-ray burst immediately diagnoses the accretor as a neutron star. For reviews, see [10] and [11]. The BeppoSAX WFCs detected 2213 X-ray bursts from 54 LMXBs (not counting the Rapid Burster and Bursting Pulsar). 23 new bursters were discovered, adding at the time 50% to the population[1]. The most prolific burster is GX 354–0 with 439 bursts. On the other end of the spectrum there are 13 sources which exhibited only a single burst.

Figure 4 presents a histogram of the burst durations. It is possible to distinguish 4 groups. Most X-ray bursts have e-folding decay times of around 5 sec, for instance all bursts from GX 354–0. They are thought to be due to flashes of pure helium when the accretion rate is so high that the accreted hydrogen burns in a stable manner to helium, creating a pure helium layer which flashes unstably every few hours (e.g., [12, 13]). The second group makes up the tail of the main peak in the duration histogram. These concern mixed helium/hydrogen flashes which occur when either there is no continuous stable hydrogen burning or when the supply of fresh hydrogen is so fast that the stable burning reaction cannot keep up. The third group concerns the 'superbursts' which are 10^3 more energetic and last for a number of hours (beyond 2000 s in the histogram). These are though to be due to shell flashes of thick carbon layers [14, 15] and were discovered with the WFCs [16]. The fourth group is made of 'intermediate bursts' that have durations between the large group of mixed helium/hydrogen bursts and the

---

[1] Currently, 80 X-ray bursters are known; see http://www.sron.nl/~jeanz/bursterlist.html

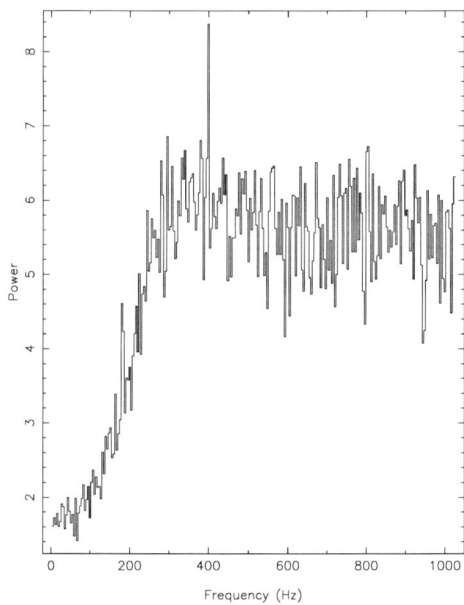

FIGURE 5. Fourier spectrum of part of an X-ray burst of SAX J1808.4-3658.

superbursts (between 80 and 600 s in the histogram). They last a good portion of an hour. These have been singled out just recently and are thought to result from very thick pure helium flashes [17, 18]. The reason that the ignition takes much longer to develop than in the short pure helium flashes, resulting in a much thicker accumulated helium layer, is probably that the accretion rate is lower so that the temperature in the helium layer is not as high and it takes a higher pressure (ergo, thicker helium pile) to reach ignition conditions. Intermediate bursts might preferentially occur in ultracompact X-ray binaries where there is no hydrogen and where the accretion rates appear to be low for prolonged periods of time (like in 2S 0918-549 and 4U 0614+091). Possibly, long helium flashes may be a good tracer of ultracompact X-ray binaries (in 't Zand et al., in prep).

The sensitivity of the BeppoSAX WFCs with respect to burst oscillations was limited. The light collecting area was 9 to 46 times less than for the RXTE Proportional Counter Array (depending on how many of the 5 units in the array are operating) with which most burst oscillations were detected. Still, it has been possible to detect an oscillation in the accretion-powered millisecond X-ray pulsar SAX J1808.4-3658, see Fig. 5 [19]. It was detected in an X-ray burst with a peak flux 9 times the flux of the Crab. This was the brightest burst seen with the WFC, although it only illuminated 50 cm^2 of all 140 cm^2 available detector area. Therefore, such an oscillation would be detectable for an on-axis burst if it is brighter than 3 Crab units. For MIRAX the timing capabilities will be significantly enhanced, thus improving the performance towards burst oscillations.

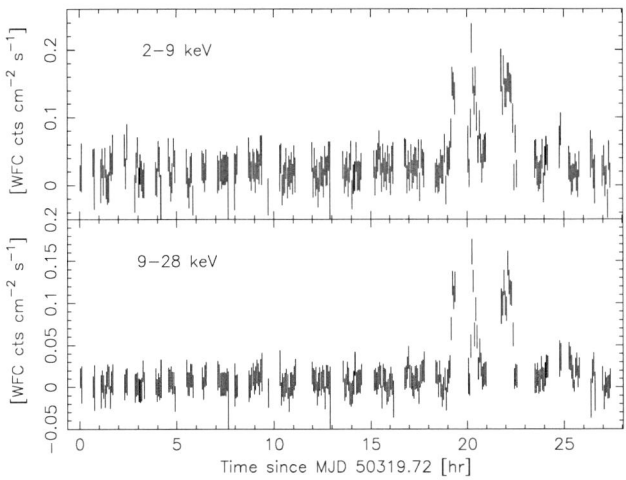

FIGURE 6. Example light curves of an outburst from the supergiant FXT IGR J17544-2619, obtained with BeppoSAX WFC in 1996.

FAST X-RAY TRANSIENTS

Fast X-ray transients are defined to have durations less than 1 day (for a review, see [20]). They form a diverse group:

Magnetic flares. These are the more energetic cousins of solar flares and occur in late-type stars with an interior convection zone. If two such stars are in a close binary the flares can be significantly more energetic. The energy is though to be released through magnetic reconnection, partly in the form of kinetic energy and partly in radiation. The WFCs detected 19 such flares from 19 different stars of types BY Dra, RS CVn, dwarf M stars and pre-main sequence stars.

Supergiant FXTs. This subclass was suspected since the late nineties [21, 22] and confirmed through observations with INTEGRAL (for a review, see [23]) and BeppoSAX [24]. BeppoSAX-WFC detected multiple outbursts from 4 supergiant systems (e.g., [24]; [25]; see Fig. 6). Currently 9 supergiant FXTs have been identified, some of them only tentatively. About half are optically identified, sometimes through accurate localizations with Chandra. All those identified involve early-type supergiants. An idea is that the short outbursts result from inhomogeneities in the wind of the supergiant that are caught by the compact object. Such structures have been suspected in single supergiants through modelling of spectral lines. Also, there is ample theoretical basis for the presence of such inhomogeneities (e.g., [26]). The X-ray outbursts may represent the strongest empirical evidence for the inhomogeneities. Further progress will come from dedicated optical spectroscopy studies (to infer wind characteristics and the binary orbit).

Gamma-ray bursts and X-ray flashes. BeppoSAX-WFC detected 89 gamma-ray bursts and X-ray flashes. 53 were identified in near real time and most of those were followed up. The remaining cases were found in the data much later, mostly because they were not detected with the concurrent Gamma-Ray Burst Monitor on BeppoSAX, and were not published. Every BATSE burst that could conclusive be localted within the field of view of the WFCs was detected, testifying to a very efficient WFC detection capability for this phenomenon (cf, [27]). The average rate of GRBs/XRFs per WFC field of view is once per 22 days. It is important to point out that WFC excels in the detection of X-ray flashes that may not be obvious in γ−ray imagers. In fact, the X-ray flash phenomenon was established as an important one through the WFC observations. Finally, WFC has the potential to detect narrow spectral features in the prompt X-ray emission [28].

PROSPECTS FOR MIRAX

The main scientific goal of MIRAX is the nearly continuous (9 months per year) broad-band (2-200 keV) high-resolution (5-7$'$) monitoring of a specific large region of the sky, the Galactic Bulge, that is particularly rich of X-ray sources [2]. The capabilities of the WFC FSM and the Hard X-ray Imager (HXI; Rothschild et al., these proceedings) are well matched for this. The field of view of the WFC FSM is completely encompassed by that of the HXI, so that all contained sources will have coverage up to 200 keV. The WFC FSM is an excellent instrument for the above-mentioned MIRAX goal, as is proven by the Galactic Bulge program on BeppoSAX as discussed here. Therefore, no hardware changes are foreseen concerning the imaging design.

The uncommon feature of the WFC FSM is that it is an 'old' instrument whose detector cannot be refurbished. The detector was sealed in 1995 which may pre-date the launch by at least 15 years. The age does not preclude a flight for several reasons: 1) the two WFCs that flew on BeppoSAX were fully functional with no detector degradation until the stop of operations in 2002 while the equally qualified FSM experienced negligible load; 2) the pre-flight performance of all the three WFCs is reproducible between each other; 3) the technology does not include expendables. Currently, the FSM is in the preliminary stages of compatibility reviews, acceptance tests and possible updates.

At least one change is foreseen for the instrument on MIRAX. The onboard software will be adjusted so that each photon is labelled with a time stamp of resolution 1/4096 s. In BeppoSAX only every 4th photon was labelled with a stamp of resolution 1/2048 s (except for a sporadic observation in a different data mode) and the average accuracy for the intermediate photons was $4/R$ if R is the total photon rate (for the Galactic center observations $4/R \approx 5$ ms). Therefore, the improvement for MIRAX to detect ms signals would be significant. Simulations and predictions show that the detection threshold for detecting ms burst oscillations in Galactic center observations is about 50% fractional rms for an on-axis burst for a 1-s times series at 1 Crab. For an off-axis position such that half the detector is used this becomes 50% worse. For a 4 Crab flux on-axis it is about 10% rms. The implied bitrate would be, for the Galactic Bulge, on average 35 kbps and peaking (for the processor limit, for instance during X-ray bursts) at 70 kbps.

For a baseline approach to the observation schedule (9 months during each of 5 years on the Galactic bulge) that is foreseen for MIRAX, we anticipate the following prospects from the WFC FSM on MIRAX:

- ∼20,000 X-ray bursts and a few tens of superbursts. The order of magnitude enhancement in exposure time with respect to BeppoSAX will 1) provide better measurements on burst recurrence times and burst quenching after superbursts (see

Cumming, these proceedings) and 2) provide more detections of rare kinds of X-ray bursts such as those of neutron stars with low accretion levels as in ultracompacts;
- detections of normal bursts from nearby SGR (WFC detected such bursts from SGR 1900+14);
- ∼40 GRBs and XRFs. The good X-ray response for these may bring about further examples of narrow absorption features in the prompt emission;
- ∼40 X-ray transients;
- incidental discovery of a burst oscillation and pulsar spin period.

ACKNOWLEDGMENTS

We thank the local organizing committee for the generous hospitality, and Gerrit Wiersma and Jaap Schuurmans for support in the data analysis.

REFERENCES

1. J.A.M. Bleeker, *Nucl. Phys. B Proc. Suppl.*, **132**, 10 (2004)
2. J. Braga, R. Rothschild, J. Heise, et al., *Adv. Sp. Res.* **34**, 2657 (2004)
3. W.A. Mels, H.B. Buurmans, R. Jager, et al., *SPIE*, **2517**, 269 (1995)
4. R. Jager, W.A. Mels, A.C. Brinkman, et al., *A&AS* **125**, 557 (1997)
5. J.J.M. in 't Zand, J. Heise, & R. Jager, *A&A* **288**, 665 (1994)
6. A.P. Hammersley, Ph.D. thesis, Univ. Birmingham, UK (1986)
7. A. Goldwurm, P. David, L. Foschini et al., *A&A*, **411**, L223 (2003)
8. J.J.M. in 't Zand, F. Verbunt, J. Heise, et al., *Nucl. Phys. B Proc. Suppl.* **132**, 486 (2004)
9. J.-P. Lasota, *NewAR*, **45**, 449 (2001)
10. W.H.G. Lewin, J. van Paradijs, & R. Taam, *Space Sci. Rev.* **62**, 223 (1993)
11. T.E. Strohmayer, & L. Bildsten, in "Compact Stellar X-Ray Sources, edited by W.H.G. Lewin & M. van der Klis, Cambridge University Press, in press (astro-ph/0301544)
12. M.Y. Fujimoto, T. Hanawa, & S. Miyaji, *ApJ* **247**, 267 (1981)
13. L. Bildsten, in Proc. "The Many Faces of Neutron Stars", eds. A. Alpar, L. Buccheri, & J. van Paradijs (Dordrecht: Kluwer), p. 419 (1998)
14. T.E. Strohmayer & E. Brown, *ApJ*, **566**, 1045 (2002)
15. A. Cumming & L. Bildsten *ApJ* **559**, L127 (2001)
16. R. Cornelisse, J. Heise, E. Kuulkers, et al., *A&A* **357**, L21 (2000)
17. J.J.M. in 't Zand, A. Cumming, M.V. van der Sluys, et al., *A&A* **441**, 675 (2005)
18. A. Cumming, J. Macbeth, J.J.M. in 't Zand, & D. Page, *ApJ*, in press (astro-ph/0508432)
19. J.J.M. in 't Zand, R. Cornelisse, E. Kuulkers, et al., *A&A* **372**, 916 (2001)
20. J. Heise & J.J.M. in 't Zand, in 'Compact Stellar X-ray Sources', eds. W.H.G. Lewin & M. van der Klis, CUP, in press (2006)
21. D.M. Smith, D. Main, F. Marshall, et al., *ApJ*, **501**, L181
22. A. Bamba, J. Yokogawa, M. Ueno, et al., *PASJ*, **53**, 1179 (2001)
23. I. Negueruela, D.M. Smith, P. Reig, et al., in Proc. "The X-ray Universe 2005", San Lorenzo de El Escorial (Madrid, Spain), 26-30 September 2005, ESA-SP 604, in press (astro-ph/0511088)
24. J.J.M. in 't Zand, J. Heise, P. Ubertini, et al., in Proc. 5th INTEGRAL Workshop, 'the INTEGRAL Universe', ESA SP-552, eds. V. Schönfelder, G. Lichti & C. Winkler, p. 427
25. J.J.M. in 't Zand, *A&A*, **441**, L1 (2005)
26. S. Owocki,*Astrop. Sp. Sc.*, **221**, 3 (1994)
27. D. Band, *ApJ*, **588**, 945 (2003)
28. L. Amati, F. Frontera, M. Vietri, et al., *Science*, 290, 953 (2000)

SCIENTIFIC SATELLITES AT INPE

Marco Antonio Chamon, Terezinha Ribeiro de Carvalho

The National Institute for Space Research (INPE)
SCIENTIFIC SATELLITES AND EQUIPMENTS COORDINATION (SCE)
São José dos Campos, São Paulo
Brazil

Abstract. General activities developed by the Brazilian Institute for Space Research in space science and technology domain are summarized. Particularly, activities in space scientific missions, where scientific experiments are developed and embarked on satellite to operate in orbit are object of this paper and are described below.

Keywords: INPE, Scientific Satellites, EQUARS, MIRAX, MCE

A BRIEF OF SPACE ACTIVITIES IN BRAZIL

The National Institute for Space Research (INPE), in Brazil, is a research unit of the Ministry of Science and Technology (MCT). The Institute has about 1200 employers spread over 10 work centers. INPE Directorate is settled in São José dos Campos-SP.

INPE started its activities by stimulating, coordinating, and supporting studies on the space related area, besides breeding a team of skilled researchers and establishing cooperation with leading nations on the space area.

At start, the research program was closely related with studies in the field of space and atmospheric sciences. Gradually, the usage of meteorological, communications and earth observation satellites emerged as the most important activities of the Institute concerning Brazilian needs.

On the late seventies, INPE entered a new era when the Federal Government approved the Complete Brazilian Space Mission (MECB); the institute, besides research and applications, started the development of space technology. While Brazil benefited from services of foreign satellites, it became clear that in a country of continental dimensions with immense and almost inhabited areas, it was essential to develop its own space technology toward its specific needs.

During the eighties, INPE started developing priority programs such as the Complete Brazilian Space Mission (MECB), China-Brazil Earth Resources Satellite (CBERS), Amazônia Research Program (AMZ) and the Center for Weather Forecast and Climatic Studies (CPTEC). It also kept track of other countries' research on the space area, facilitating collaboration and partnership with them. During this period it also established some facilities like Integration and Tests Laboratory (LIT), the Satellite Control Center and Ground Stations.

In 1993, the first Brazilian data collecting satellite (SCD-1) was launched showing the Brazilian competences in space technology. In 1998, the second Brazilian satellite (SCD-2) was successfully launched, performing even better than the first one.

The China-Brazil Earth Resources Satellite (CBERS) Program began in 1988 with an agreement between Brazil and China to produce and launch two remote sensing satellites. The success of CBERS-1 launched in 1999 and the CBERS 2 launched in 2003, and the perfect operation of the satellites generated immediate effects, both governments decided to expand the initial agreement by including another two satellites of the same kind, CBERS-3 and 4, as the second stage of the Sino Brazilian cooperation effort.

Recently, a proposal for developing an university micro-satellite, with strong participation of student from Technological Institute of Aeronautics (ITA) was approved. By the proposal INPE will provide management and technical support for designing, manufacturing and testing a technological mission where new developments for space platforms are implemented and tested in orbit.

Activities at INPE in space scientific missions, where scientific experiments are developed and embarked on satellite to operate in orbit are described in next section.

SPACE SCIENTIFIC MISSIONS AT INPE

This section describes the main activities concerning scientific satellites, in charge of the Scientific Satellites and Equipments Coordination (SCE). As a division of the Space Engineering and Technology Coordination (ETE) is engaged in developing experiments for space missions and scientific satellites.

For the near future INPE will develop and launch several satellites for research and scientific explorations. The proposed scientific missions EQUARS, MIRAX and MCE are described below.

Equatorial Atmosphere Research Satellite (EQUARS)

EQUARS's objectives are to understand atmospheric coupling between dynamical, electrical, photochemical and ionospheric processes, and to apply the data to atmospheric, space weather and climate studies.

The topics to be investigated in equatorial atmosphere monitoring are: a) troposphere: water vapor profile; cloud convection and lightning; b) stratosphere: temperature variability; c) mesosphere: wave propagation; temperature variability; d) ionosphere: generation and propagation of plasma bubbles.

The results will be applied in real-time monitoring of the tropospheric water vapor, stratospheric temperature profile and total electron contents (TEC) in the ionosphere by the GPS occultation measurements. These results will make possible to carry on high quality numerical weather and climate predictions and space weather forecasting. In order to achieve it, close collaborative operation with the COSMIC project is proposed.

Satellite Characteristics

Orbit	Equatorial, Altitude 750 km (LEO), Inclination 20 deg
Total Mass	under 100 kg
Volume	60 x 70 x 80 (in cm)
Power Consumption	140 W
Attitude Control	Active, 3-axis stabilized
Orientation	Geocentric (1 deg precision)
Data Transmission	1-2 Mbps and two ground stations: one at Alcântara (northeast Brazil) and another in the Asian region (under study)

Monitor e Imageador of Raios-X (MIRAX)

MIRAX is an X-ray astronomy satellite mission presently in development by the high energy astrophysics group of the Astrophysics Division (DAS) of INPE in Brazil. The main scientific goals of MIRAX are based on a unique capability of the mission: continuous viewing (at least for 9 months) of a rich region of X-ray sources. The fully coded field-of-view (FCFOV) for the hard X-ray imager will be 39° x 6°, centered on the Galactic Center and with the longer axis aligned with the Galactic Plane. This will not only provide an unprecedented monitoring of the X-ray sky through simultaneous spectral observations of a large number of sources, but will also allow the detection, localization, possible identification, and spectral/temporal study of the entire history of transient phenomena to be carried out in one single mission. MIRAX will have sensitivities of < 5 mCrab/day in the 2-10 keV band (~ 2 times better than the All Sky Monitor on RXTE) and 2.6 mCrab/day in the 10-100 keV band (~ 40 times better than the Earth Occultation technique of the Burst and Transient Source Experiment on the Compton Gamma Ray Observatory).

The MIRAX mission is expected to provide significant scientific contributions on the following topics:

- Complete history of transient sources
- Accretion torques on neutron stars through observations of X-ray pulsars and burst oscillations
- Spectral state changes and evolution on black-hole systems
- Relativistic jets on microquasars (X-ray light curves during radio ejections)
- Fast X-ray novae, X-ray bursts, Soft Gamma-ray Repeaters
- Gamma-ray bursts (~1/month)

Satellite Characteristics

Orbit	Equatorial, Altitude 600 km (LEO), Inclination 0 deg	
Total Mass	250 kg (total), 125 kg (payload)	
Power Consumption	240 W (total), 90 W (payload)	
Attitude Control	Active, 3-axis stabilized	
Data Transmission	S-band (2200-2290 MHz), 1.5 Mbps downlink; one ground station at Alcântara (northeast Brazil)	
Instrument parameters	Hard X-ray Imager	Soft X-ray Imager
energy range	10-200 keV	2-30 keV

Space Weather Monitor (MCE)

This mission comprehends a constellation of 3 satellites for space climate monitoring in an international cooperation between Russia / Ukraine / Brazil (IKI / Univ. Kiev / INPE). The mission are still in pre-phase A for feasibility analysis.

The MCE mission is expected to provide significant scientific contributions on the following topics:

- Space climate monitoring and forecasting
- Solar wind-magnetosphere coupling analysis through multiple observations
- Solar flares warning and forecasting

CONCLUDING REMARKS

Brazilian space program has focused its development on low Earth orbit (LEO) satellites for meteorological data collecting, remote sensing, and scientific missions. Particularly, Brazilian scientific satellites call for a large international cooperation not only in terms of scientific data exploitation but also as an opportunity for international on-board experiments, providing Brazilian global insertion in space science and technology.

FURTHER INFORMATION

National Institute for Space Research (INPE) - http://www.inpe.br/english

Ministry of Science and Technology (MCT) - http://www.mct.gov.br (only in Portuguese)

Brazilian Space Mission (MECB) - http://www.inpe.br/programas/mecb/ingl/default.htm

China-Brazil Earth Resources Satellite (CBERS) - http://www.cbers.inpe.br

University Satellite (ITASAT) - http://www.itasat.ita.br
EQUARS - http://www.laser.inpe.br/equars/eng/siteEquars.shtml
MIRAX - http://www.cea.inpe.br/cea/satelites/mirax/miraxproject.htm
MCE - http://www.cea.inpe.br/cea/satelites/mce (only in Portuguese)

MIRAX SCIENCE—X-RAY

IGR Sources and *MIRAX*

John A. Tomsick

Center for Astrophysics and Space Sciences, Code 0424, University of California at San Diego, La Jolla, CA 92093, USA

Abstract.
As the *INTEGRAL* mission increases its exposure time of the Galactic plane, it is giving us a much more complete picture of hard X-ray sources in the Galaxy, including the discovery of many new "IGR" sources. The fraction of new sources that are High-Mass X-ray Binaries (HMXBs) is large when compared to the source types for previously known sources, and several of these new HMXBs have interesting and extreme properties. In particular, some of the HMXBs exhibit bright, short flares, and using the *INTEGRAL* results as a guide, we estimate that >100 of these sources may be discovered by *MIRAX* if the *MIRAX* field-of-view covers the inner spiral arms of the Galaxy. Another benefit of covering regions with large numbers of HMXBs is that *MIRAX* will be capable of detecting pulsations from these sources, and its long term monitoring will provide orbital measurements.

Keywords: X-ray sources, instrumentation
PACS: 97.80, 07.85

INTRODUCTION

With the *MIRAX* mission focusing on X-ray imaging of the Galactic plane, it is essential to consider what recent X-ray missions are finding with Galactic plane monitoring. One of the key missions in this respect is the *INTErnational Gamma-RAy Laboratory (INTEGRAL)* mission [1], which launched in 2002 October. With large field of view ($\sim$100 deg^2) instruments and observing programs that emphasize coverage of the Galactic plane such as the "Galactic Plane Scans" (GPS) and the "Galactic Center Deep Exposure," *INTEGRAL* is providing a wealth of information about Galactic sources (also see the contribution to these proceedings by R. Walter). With these attributes, along with the ISGRI instrument's several arcminute angular resolution and energy coverage above 20 keV, *INTEGRAL* is finding a large number of new (or previously poorly studied) "IGR" sources. Figure 1 shows an example of the types of images that *INTEGRAL* is producing.

In this work, we compare the newly discovered IGR sources to the population of previously known hard X-ray sources in the Galactic plane to see how our understanding of the hard X-ray sky is changing. Then, we discuss implications for the *MIRAX* mission, which will certainly observe this population of sources and will also discover its own new X-ray sources.

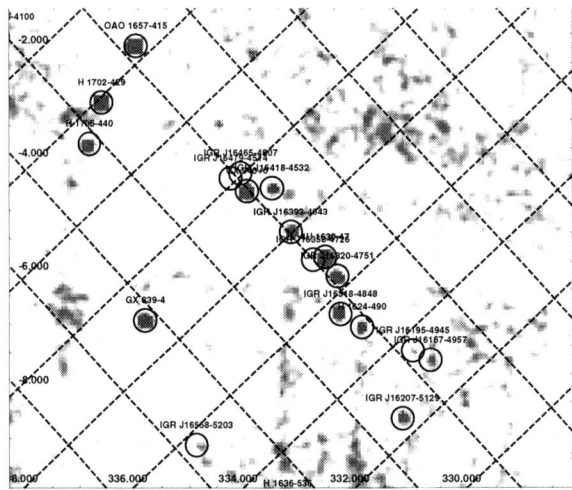

FIGURE 1. 20–40 keV *INTEGRAL*/ISGRI image of the Norma spiral arm region. This image includes an exposure time of 300 ks. All the IGR sources that have been discovered in this region are marked, but not all of them were detected in this particular image.

IGR SOURCES VS. PREVIOUSLY KNOWN SOURCES

We used two lists of hard X-ray sources for our comparison of IGR sources to previously known hard X-ray sources. First, we used the 2nd *INTEGRAL* catalog [2], which includes both new and previously known sources from the first ∼2 years of the *INTEGRAL* mission. This catalog includes sources from >10 Ms of exposure time reaching a sensitivity of ∼1 millicrab with ∼50% sky coverage. Second, the list of new sources discovered by *INTEGRAL* is constantly updated at the "*INTEGRAL* Sources" website[1], and we took the sources on this list as of 2005 November. We merged these two lists, and we refer to the sources from the website as the IGR sources. In all, our sample includes 247 sources that have been detected by *INTEGRAL*: 101 IGR sources and 146 "known" sources.

We compared the fluxes, spatial distributions, and types of sources for the IGR and known sources. Not surprisingly, the known sources are brighter, but it is not immediately clear if this is because the IGR sources are simply more distant or if they have intrinsic differences. The spatial distributions have similarities, but the known sources are somewhat more peaked in Galactic latitude (b) and longitude (l). The IGR l distribution has a strong peak in the Norma Region of the Galaxy, at $l = 330$–$340°$, indicating that many new sources are being found here (see Figure 1). About half of the IGR sources are still of unidentified nature, but 48 of the IGR sources have been identified as 21 High-Mass X-Ray Binaries (HMXBs), 8 Low-Mass X-Ray Binaries

---

[1] http://isdc.unige.ch/ rodrigue/html/igrsources.html

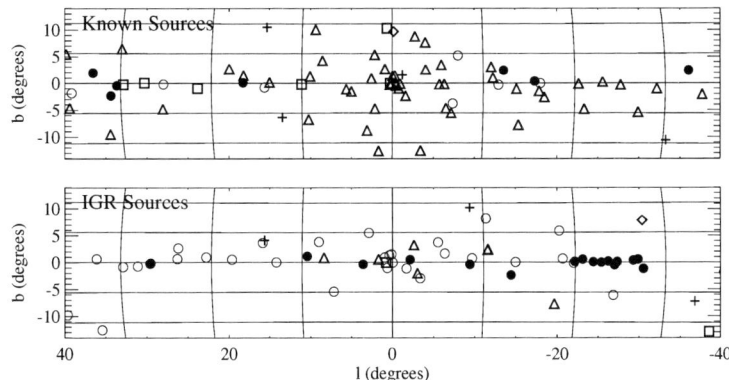

FIGURE 2. Types of known (top) and IGR (bottom) sources in the (extended) Galactic center region. the axes are Glactic longitude (l) and latitude (b). In both panels, the symbols correspond to the following: Unidentified (open circles); HMXBs (closed circles); LMXBs (triangles); AGN (pluses); CVs (diamonds); Other types (squares).

(LMXBs), 13 Active Galactic Nuclei (AGN), 3 Cataclysmic Variables (CVs), and 3 that do not fall into these catagories. The numbers of the various types for the known sources are: 31 HMXBs, 65 LMXBs, 26 AGN, 5 CVs, and 11 of other types. The striking difference is that 44% of the IGR sources that have been classified are HMXBs compared to only 23% of the known sources. On the other hand, the known sources include mostly (65%) LMXBs compared to only 17% for IGR sources. This difference is illustrated in Figure 2, and is likely due to the fact that the LMXBs are quite bright in the soft X-ray band and so they have been previously discovered by other satellites. However, the HMXBs tend to be fainter in the soft X-ray band as they are harder X-ray sources, and it took a relatively sensitive hard X-ray imager to find them.

From follow-up observations of IGR sources in the X-ray, optical, and IR, we have learned a significant amount about these sources. It has been shown that many of the sources are extreme in one way or another. Some of the sources have extremely high column densities ($N_H \sim 10^{23-24}$ cm^{-2}), and, in some cases, it has been shown that this is due to the interaction of the compact object with a strong stellar wind [3]. Other sources have been shown to be accreting X-ray pulsars, and extremely long pulse periods (1000–6000 s) are seen in some cases (see the paper in these proceedings by R. Walter for more specifics). There are three main reasons why many of the IGR sources were not found before *INTEGRAL*. First, high column densities caused absorption that kept them from being detected. A second reason is that some of the sources have very hard spectra even though their column densities are low. IGR J17195–4100 and IGR J16167–4957 are two examples of sources for which this is the case (Tomsick et al., in prep.). Third, many of the IGR HMXBs are only bright enough to be detected when they exhibit brief and bright flares [4, 5, 6]. Based on a sample of 9 flaring sources, it has been suggested that these sources may comprise a new group of "Supergiant Fast X-ray Transients" (SFXTs) with ∼100 millicrab flares that last for ∼4 hours [7].

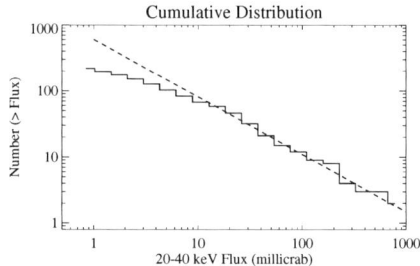

FIGURE 3. Cumulative flux distribution for *INTEGRAL*-detected sources (IGR and "known"). The dashed line suggests that the survey may be nearly complete down to ∼10 millicrab. If the survey is ultimately complete to 0.5–1 mimllicrab, 500–1000 sources could ultimately be found.

IMPLICATIONS FOR *MIRAX*

The sources and the behavior of the sources detected by *INTEGRAL* provide very useful information for the *MIRAX* mission. Figure 3 shows the cumulative distribution of *INTEGRAL*-detected sources. The distribution indicates that the *INTEGRAL* surveys may currently be complete to about 10 millicrab, and that, if the survey is ultimately complete to 0.5–1.0 millicrabs, a total of 500–1000 hard X-ray sources will be detected. With its continuous coverage (other than Earth occultations) of the Galactic center region, *MIRAX* will allow us to study these sources in hard and soft X-rays over a very large range of time scales, and will provide an excellent opportunity to find, e.g., pulsations, state transitions, X-ray flares, X-ray bursts, etc. We have performed simulations to determine the *MIRAX* capabilities for studying accreting X-ray pulsars. With its hard X-ray coverage, in 2 days of observations, *MIRAX* will be capable of measuring X-ray pulsations for sources as faint as 5 millicrab. Thus, the standard *MIRAX* observing program will provide orbital solutions (via measurements of pulsations over time) for large numbers of faint X-ray pulsars.

The behavior of the IGR sources also indicates that *MIRAX* should find large numbers of new hard X-ray sources. Due to its observing strategy, *MIRAX* will excel at finding flaring hard X-ray sources and fast transients (see the contribution to these proceedings by R. Rothschild for a figure showing the *MIRAX* "discovery space"). Here, we use the properties of the SFXTs and the *INTEGRAL* observing strategy to estimate the numbers of SFXTs that *MIRAX* might find (see Table 1). In some cases, flares from SFXTs repeat, and the first column of Table 1 gives the number of outbursts per source assumed. The second column gives the assumed duration of the outburst. Given the flare properties in the first 2 columns, columns 3 and 4 give the probability that the source would be detected in a year of observing with the *INTEGRAL* Galactic Plane Scans and with the nominal *MIRAX* observing strategy (9 months per year on the Galactic center region with only Earth occultations). The final column gives the estimate for the number of *MIRAX* detections expected per *INTEGRAL* detection. For the 4 hour bursts, the numbers indicate that *MIRAX* will be 44–100 times more sensitive than the *INTEGRAL*/GPS. Although the 9 SFXTs discussed above were found in various ways, we estimate that

TABLE 1. Estimates of how many new fast transients *MIRAX* might find.

Number of Outbursts per source per year	Durations of Outbursts	Detection Probability per year *INTEGRAL*/GPS	Detection Probability per year *MIRAX*	Number of *MIRAX* detections per one *INTEGRAL* detection
1	4 hrs.	0.75%	75%	100
2	4 hrs.	1.49%	93.8%	63
3	4 hrs.	2.23%	98.4%	44
1	0.4 hrs.	0.2%	65%	325

$\sim$2 were found during the *INTEGRAL*/GPS, indicating that $>$100 of these sources might be found by *MIRAX*. It is also notable that the *MIRAX* capabilities are more impressive for even shorter bursts (if they exist), and the numbers for the 0.4 hour case are given in Table 1.

While *MIRAX* is likely to find many new flaring sources and fast transients, it is important to note that the above calculations assume that the bursting sources are in the *MIRAX* field-of-view (FOV). While *MIRAX* science has primarily been focused on the Galactic center region, the hard X-ray sources such as the SFXTs and the HMXB accreting X-ray pulsars tend to be in the spiral arms such as the Norma ($l = 330$–$340°$) and the Scutum ($l = 20$–$30°$) arms. Thus, for *MIRAX* to be optimized to perform this science, its FOV must cover at least these inner spiral arms.

ACKNOWLEDGMENTS

JAT thanks the organizers (scientific and local) of the workshop for planning and conducting the workshop, and a special thanks to Joao Braga and Flavio D'Amico for their hospitality.

REFERENCES

1. C. Winkler, T. J.-L. Courvoisier, G. Di Cocco, N. Gehrels, A. Giménez, S. Grebenev, W. Hermsen, J. M. Mas-Hesse, F. Lebrun, N. Lund, G. G. C. Palumbo, J. Paul, J.-P. Roques, H. Schnopper, V. Schönfelder, R. Sunyaev, B. Teegarden, P. Ubertini, G. Vedrenne, and A. J. Dean, *A&A* **411**, L1–L6 (2003).
2. A. J. Bird, E. J. Barlow, L. Bassani, A. Bazzano, G. Bélanger, A. Bodaghee, F. Capitanio, A. J. Dean, M. Fiocchi, A. B. Hill, F. Lebrun, A. Malizia, J. M. Mas-Hesse, M. Molina, L. Moran, M. Renaud, V. Sguera, S. E. Shaw, J. B. Stephen, R. Terrier, P. Ubertini, R. Walter, D. R. Willis, and C. Winkler, *ApJ* **636**, 765–776 (2006).
3. P. Filliatre, and S. Chaty, *ApJ* **616**, 469–484 (2004).
4. D. M. Smith, D. Main, F. Marshall, J. Swank, W. A. Heindl, M. Leventhal, J. J. M. in 't Zand, and J. Heise, *ApJ* **501**, L181+ (1998).
5. J. J. M. in't Zand, *A&A* **441**, L1–L4 (2005).
6. V. Sguera, E. J. Barlow, A. J. Bird, D. J. Clark, A. J. Dean, A. B. Hill, L. Moran, S. E. Shaw, D. R. Willis, A. Bazzano, P. Ubertini, and A. Malizia, *A&A* **444**, 221–231 (2005).
7. I. Negueruela, D. M. Smith, P. Reig, S. Chaty, and J. M. Torrejon, *ArXiv Astrophysics e-prints* (2005), arXiv:astro-ph/0511088.

The INTEGRAL Galactic Bulge monitoring program

E. Kuulkers[*], S.E. Shaw[†,**], S. Brandt[‡], J. Chenevez[‡], T.J.-L. Courvoisier[**], K. Ebisawa[§], P. Kretschmar[*], C.B. Markwardt[¶,||], N. Mowlavi[**], T. Oosterbroek[††], A. Orr[††], A. Paizis[‡‡], C. Sanchez-Fernandez[*] and R. Wijnands[§§]

*ISOC, ESAC/ESA, Apartado 50727, 28080 Madrid, Spain
†University of Southampton, UK
**ISDC, Switzerland
‡DNSC, Denmark
§ISAS, Japan
¶University of Maryland, USA
||NASA/GSFC, USA
††ESA-ESTEC, The Netherlands
‡‡INAF-IASF, Italy
§§University of Amsterdam, The Netherlands

Abstract. The Galactic Bulge region is a rich host of variable high-energy point sources. These sources include bright and relatively faint X-ray transients, X-ray bursters, persistent neutron star and black-hole candidate binaries, X-ray pulsars, etc.. We have a program to monitor the Galactic Bulge region regularly and frequently with the γ-ray observatory *INTEGRAL*. As a service to the scientific community the high-energy light curves of all the active sources as well as images of the region are made available through the WWW. We show the first results of this exciting new program.

Keywords: Accretion and accretion disks; Neutron stars; Black holes; X-ray binaries; Galactic center and bulge; X-ray sources; X-ray bursts; gamma-ray sources
PACS: 97.10.Gz; 97.60.Jd; 97.60.Lf; 97.80.Jp; 98.35.Jk; 98.70.Qy; 98.70.Rz

INTRODUCTION

The bulge of our Galaxy hosts a variety of X-ray and γ-ray point sources (e.g., Knight et al. 1985, Skinner et al. 1987, Churazov et al. 1994; see, e.g., Bird et al. 2006, Bélanger et al. 2006, Revnivtsev et al. 2004, for *INTEGRAL* observations). These include persistent and transient neutron star and black-hole candidate binaries, X-ray pulsars, X-ray bursters, etc.. Because of the variability these sources possess on time scales of msec to days (quasi-periodic oscillations, pulsations, [absorption] dips, eclipses, type I and type II X-ray bursts, orbital variations, flares) and weeks to years (orbital variations, outburst cycles, on/off states), the region never looks exactly the same.

From 17 February 2005 onwards *INTEGRAL* has been monitoring this region approximately every 3 days, as part of our approved AO-3 program, whenever the region is visible by *INTEGRAL*. In this paper we describe this program in more detail and show the first results of the first two periods of monitoring performed between February and October 2005.

INTEGRAL AND DATA ANALYSIS

INTEGRAL (The **Inter**national **G**amma-**R**ay **A**strophysics **L**aboratory; Winkler et al. 2003) is an ESA scientific mission dedicated to fine spectroscopy ($E/\Delta E \simeq 500$; SPI) and fine imaging (angular resolution: 12 arcmin FWHM; IBIS) of celestial γ-ray sources in the energy range 15 keV to 10 MeV with simultaneous monitoring in the X-ray (3–35 keV; JEM-X) and optical (V-band, 550 nm; OMC) energy ranges.

Our program is to observe the region frequently and regularly, with the aim to investigate the source variability and transient activity on time scales of days to weeks to months at relatively soft ($\lesssim 10$ keV) and hard ($\gtrsim 10$ keV) energies. One complete hexagonal dither pattern (7 pointings of 1800 sec each, i.e., 1 on-axis pointing, 6 off-source pointings in a hexagonal pattern around the nominal target location, each 2° apart) is performed during each *INTEGRAL* revolution, or orbit around the earth (i.e., roughly every 3 days). This is done whenever the region is visible by *INTEGRAL* (about two times per year for a total period of about 4 months). As a service to the scientific community, the JEM-X light curves (3–10 keV and 10–25 keV) and the IBIS/ISGRI light curves (20–60 keV and 60–150 keV) are made publicly available as soon as possible after the observations are performed. In addition, IBIS/ISGRI and JEM-X mosaic images of each hexagonal observation are provided, with information on the detected sources. Last, but not least, all IBIS/ISGRI 20–60 keV mosaic images per revolution are stacked into a movie, showing the ever-changing gamma-ray sky. All the instruments onboard *INTEGRAL*, except the OMC, have coded masks. With the fully and partially coded field of views (FOVs) we cover about half of the low-mass X-ray binary (LMXB) and high mass X-ray binary (HMXB) Galactic Bulge population (see also, e.g., in 't Zand 2001).

Similar Galactic Bulge monitoring programs have been performed (see, e.g., in 't Zand 2001) and are currently ongoing (e.g., *RXTE* Galactic Bulge Scans, Swank & Markwardt 2001). However, the *RXTE*/PCA and HEXTE do only have a 2° collimator, so only a small field of view with no imaging resolution, and therefore only provide information on a given source for a short time when the instrument scans over it; moreover, in the Galactic Center region itself there is some source confusion. There are currently other instruments in operation at similar energy ranges (e.g., *Swift*/BAT: 15–150 keV with a FOV of 2 steradians; Barthelmy 2000), but they do not provide frequent and regular monitoring of the Galactic Bulge region, as well as having a worse imaging capability, again leading to some source confusion in the Galactic Center region (*Swift*/BAT PSF angular resolution is 22' compared to the IBIS/ISGRI angular resolution of 12' [FWHM]).

For our program at the moment we only consider data from IBIS/ISGRI (Ubertini et al. 2003, Lebrun et al. 2003) and JEM-X (Lund et al. 2003). We do not consider the data from the IBIS/PICsIT, SPI, or OMC instruments. Either the angular resolution is high (SPI: 2.5°) and therefore the various sources in the Galactic Bulge region close to each other complicate the analysis, or the sources are too weak to be detected (IBIS/PICsIT). For the OMC, however, we are currently evaluating the scientific output.

The *INTEGRAL* data reduction is performed using the Off-line Scientific Analysis (OSA; Courvoisier et al. 2003), v5.1. We use a source catalog, currently containing 79 sources which have been detected by IBIS/ISGRI up to now in the field we are interested in (but see next Section).

The data from IBIS/ISGRI are processed until the production of images in the 20–60 and 60–150 keV energy ranges per single exposure. We force the flux extraction of each of the catalogue sources, regardless of the detection significance of the source. This method is essential in order to clean the images from the ghosts of all the active sources in the field, but does not make any threshold selection and all the positive fluxes are recorded. In order to detect fainter sources, we then mosaic the images from the single exposures and search for all catalog sources, as well as possible new ones. For JEM-X the analysis is run through the imaging step to the light-curve step in OSA for a single bin of the same length as the exposure window. Light curves are produced for all catalog sources up to 5° off from the center of the FOV. Again, the images from the single exposures are mosaiced in order to create the final image (but no further source detection was done).

Per hexagonal dither (i.e., 7 exposures combined) we are sensitive down to typically between 5 and 15 mCrab (6σ) for both JEM-X and IBIS/ISGRI. The actual sensitivity depends on factors such as source position (fully or partially coded FOV), background (instrument systematics, solar activity), number of exposures (some are lost) and energy (instrument response).

The results, as well as more information about the program, can be retrieved from the *INTEGRAL* Galactic Bulge Monitoring WWW page hosted at the ISDC in Switzerland: http://isdc.unige.ch/Science/BULGE/.

GALACTIC BULGE MONITORING: FIRST RESULTS

By now we have had two full seasons of monitoring, i.e., from revolutions 287–307 (2005 February - April) and 347–370 (2005 August - October), respectively. The third season started in revolution 406 (February 2006). In Figure 1 we show examples of results from the first two seasons for various types of sources. At the left we show the light curves of (temporary) bright (i.e., easily detected in one exposure) sources. At the right we show the light curves of weaker (persistent or transient) or slowly varying sources, averaged per revolution.

Whereas, for example, GX 3+1 (neutron star LMXB) does not vary much, sources like GX 1+4 (symbiotic binary containing a neutron star) and 1A 1742−294 (neutron star LMXB) vary on monthly times scales, while 1E 1740.7−2942 (LMXB) varies smoothly on even longer time scales. Some sources clearly show transient behaviour, i.e., they show outbursts with durations exceeding months (e.g., H1743−322, a black-hole candidate LMXB; see also below) or flaring on timescales of hours to days (e.g., IGR J17252−3616, an X-ray pulsar HMXB). Some sources vary on all timescales accessible through our program, as displayed by the HMXB 4U 1700−377.

Similarly, sources like 1A 1742−289 and 1E 1742.2−2857 (both unidentified X-ray sources) show low-level activity on various timescales.[1] Sources like the neutron

---

[1] These sources are not included in our source catalog, but the light curves displayed here are the result when the ISDC Reference catalog is used as input. Note that these sources were not reported by Bélanger et al. 2006 and Bird et al. 2006, based on long exposure times. Further investigation is in process.

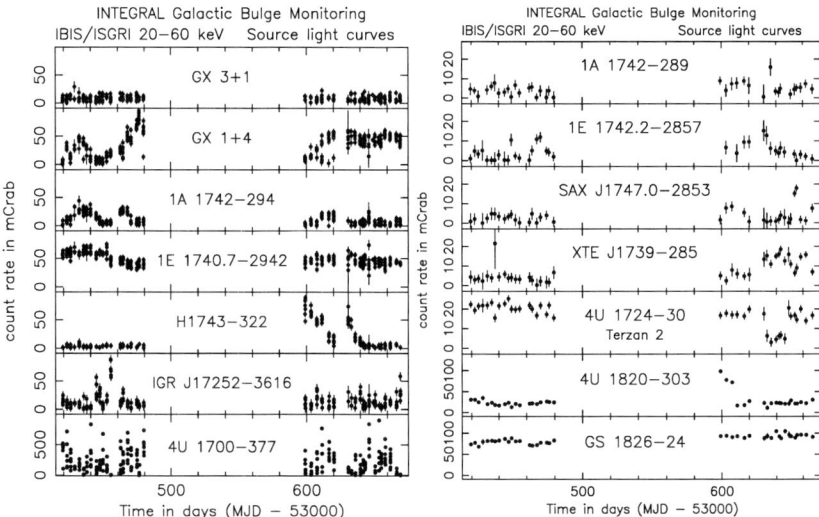

FIGURE 1. *INTEGRAL* IBIS/ISGRI (20–60 keV) light curves from the first two seasons of the Galactic Bulge monitoring program. Shown are some examples of the sources detected. The light curves are either on a time scale of one exposure, i.e., 1800 sec (*left*) or on a time scale of one revolution, i.e., roughly 3 days (*right*).

star LMXB transients SAX J1747.0−2853 and XTE J1739−285 showed renewed outburst activity, during the second season (see also below). The neutron star LMXBs 4U 1724−30 (in the globular cluster Terzan 2) and 4U 1820−303 have been persistently on through the seasons, displaying occasionally drops or flares, respectively, in intensity for about a month. The neutron star LMXB GS 1826−24 slowly varied through our observing periods.

So far quick-look results during the two seasons have been reported in 10 ATel's. Here we describe some of the highlights. Precisely at the start of the program the black-hole X-ray transient GRO J1655−40 was reported to become active (Markwardt & Swank 2005). The *INTEGRAL* GRO J1655−40 light curves (see Kuulkers et al. 2005a) nicely complement observations at soft X-ray (*RXTE*, see Homan 2005) and radio (*VLA*; see Rupen et al. 2005) wavelengths. Various other transient sources popped up and faded away, such as The Rapid Burster, H1743−322 (both Kretschmar et al. 2005; for H1743−322 see Figure 1, left), IGR J17098−3628 (Mowlavi et al. 2005), SAX J1747.0−2853 (Kuulkers et al. 2005b; see Figure 1, right) and XTE J1818−245 (Shaw et al. 2005a). In 2005 August, the X-ray transient XTE J1739−285 was found by *INTEGRAL* to be bright at soft and not detected at hard X-ray energies (Bodghee et al. 2005). About a month later the situation had reversed; it was bright at hard and weak at soft X-ray energies (Shaw et al. 2005b; see Figure 1, right). Although at first we attributed the state change to the compact object being a black hole, we proved it to be a neutron star based on the occurrence of type I X-ray bursts detected with JEM-X (Brandt et al. 2005).

CONCLUSIONS

We have shown that most of the sources in the program in the field of view of the *INTEGRAL* instruments clearly vary on timescales of a few hours to days to months; it is therefore of no surprise that the Galactic Bulge is a region to stay tuned on. MIRAX with its wide-field instruments covering a similar energy range (Braga et al. 2004; see also these Proceedings) will go a step further, i.e., it will *continuously* monitor the Galactic Bulge region for about 9 months per year down to a sensitivity level of $\simeq$5 mCrab per day. Our monitoring program is, therefore, also an ideal 'training session' for what to expect with MIRAX.

ACKNOWLEDGMENTS

Based on observations with *INTEGRAL*, an ESA project with instruments and science data centre funded by ESA member states (especially the PI countries: Denmark, France, Germany, Italy, Switzerland, Spain), Czech Republic and Poland, and with the participation of Russia and the USA.

REFERENCES

26. S. D. Barthelmy, in *Proc. SPIE* **4140**, pp. 50–63 (2000)
26. G. Bélanger, et al., *ApJ* **636**, 275–289 (2006)
26. A. J. Bird, et al., *ApJ* **636**, 765–776 (2006)
26. A. Bodghee, et al., *ATel* **#592** (2005)
26. J. Braga, et al., *Advances in Space Research* **34**, 2657–2661 (2004)
26. S. Brandt, et al., *ATel* **#622** (2005)
26. E. Churazov, et al., *ApJS* **92**, 381–385 (1994)
26. T. J.-L. Courvoisier, et al., *A&A* **411**, L53–L58 (2003)
26. J. Homan, *ATel* **#440** (2005)
26. J. in 't Zand, in *Exploring the gamma-ray universe*, editors A. Gimenez, V. Reglero, and C. Winkler, ESA SP-459, ESA Publications Division, Noordwijk, pp. 463–470 (2001)
26. F. K. Knight, et al., *ApJ* **290**, 557–567 (1985)
26. P. Kretschmar, et al., *ATel* **#593** (2005)
26. E. Kuulkers, et al., *ATel* **#438** (2005a)
26. E. Kuulkers, et al., *ATel* **#642** (2005b)
26. F. Lebrun, et al., *A&A* **411**, L141–L148 (2003)
26. N. Lund, et al., *A&A* **411**, L231–L238 (2003)
26. C. B. Markwardt, and J. H. Swank, *ATel* **#414** (2005)
26. N. Mowlavi, et al., *ATel* **#453** (2005)
26. M. G. Revnivtsev, et al., *AstL* **30**, 382–389 (2004)
26. M. P. Rupen, A. J. Mioduszewski, and V. Dhawan, *ATel* **#441** (2005)
26. S. E. Shaw, et al., *ATel* **#583** (2005a)
26. S. E. Shaw, et al., *ATel* **#615** (2005b)
26. G. K. Skinner, et al., *Nat* **330**, 544–547 (1987)
26. J. Swank, and C. Markwardt, in *ASP Conference Proceedings* **251**, edited by H. Inoue and H. Kunieda, Astronomical Society of the Pacific, San Francisco, pp. 94–97 (2001)
26. P. Ubertini, et al., *A&A* **411**, L131–L140 (2003)
26. C. Winkler, et al., *A&A* **411**, L1–L6 (2003)

An INTEGRAL view of the inner Galaxy

Roland Walter

INTEGRAL Science Data Centre, Observatoire de Genève, Chemin d'Ecogia 16, 1290 Versoix, Switzerland

Abstract. INTEGRAL is observing very regularly the galactic plane and in particular the inner regions of the Galaxy. Selected INTEGRAL results on point sources together with performance or operational aspects are discussed in relation with the MIRAX mission.

Keywords: X- and gamma- ray telescopes and instrumentation, X-ray binaries, Black-holes
PACS: 95.55.Ka, 97.80.Jp, 97.60.Lf

INTRODUCTION

The **INT**ernational **G**amma-**R**ay **A**strophysics **L**aboratory of the European Space Agency was launched on October 17, 2002. The nominal mission started on December, 30, 2002 and since then the scientific payload has worked almost flawlessly and the operations were conducted with a very high efficiency.

The two main scientific instruments consist of the imager IBIS [1] and of the spectrometer SPI [2], providing respectively sub-arcmin source positioning and keV spectral resolution in a band ranging from 17 keV to few MeV. Both instruments use coded-masks for the imaging. Substantial monitoring capabilities are also provided in the X-rays (3-35 keV) and in the optical V band by the JEM-X [3] and OMC [4] instruments. By multiplexing many observations, the large INTEGRAL field of view allows to obtain Msec effective exposure times and sub mCrab sensitivity in many areas of the sky.

Below 100 keV, the IBIS spatial resolution and improved sensitivity become key features to resolve the numerous point sources of hard X-ray emission detected in particular within the Galactic bulge and arms.

The INTEGRAL observing program is driven by selected open time observation proposals and by the guaranteed core program. In practice, when the inner galaxy is visible (this is limited by thermal constraints), most of the time is spent observing the galactic plane. INTEGRAL is therefore effectively observing any point within the MIRAX inner Galaxy field for about 10% of the time.

The INTEGRAL imager has a detector area that is about 3.5 times larger than those of the two MIRAX hard X-ray cameras. The relative strength of the background, dominated by diffuse emission, is proportional to the instrument field of view and will be twice larger for MIRAX. MIRAX will however be sensitive to slightly lower energies and spend much more exposure time on the inner galaxy increasing the rate of detected transient by a factor of 4. The flux of the faintest sources detected over the mission life time above 20 keV by MIRAX and IBIS could be roughly equivalent.

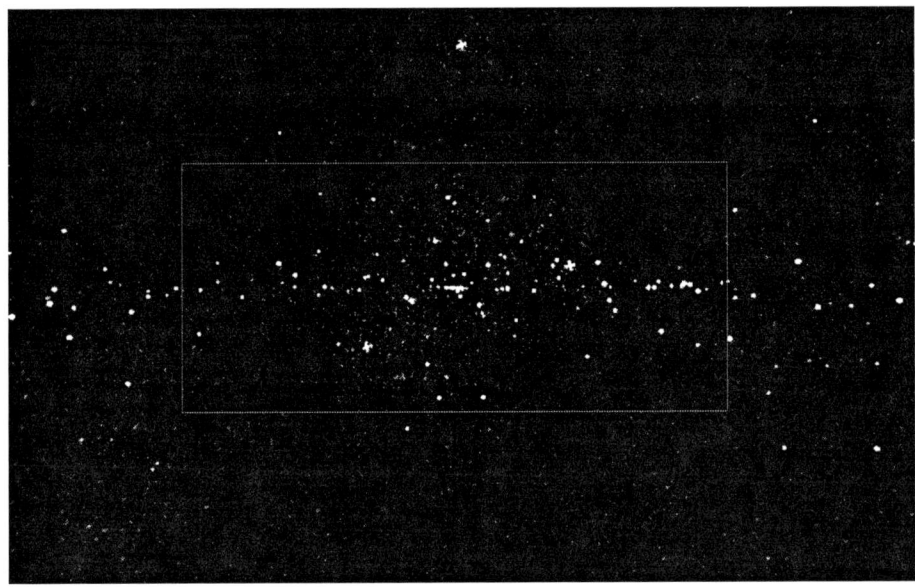

FIGURE 1. INTERGRAL/IBIS view of the inner galaxy. The MIRAX field of view is indicated.

POINT SOURCES OBSERVED BY INTEGRAL

INTEGRAL will reach an effective exposure time above 10 Msec and a sensitivity limit of the order of 0.1 mCrab between 20 and 60 keV in the central regions of the Galaxy (Fig. 1). At that sensitivity the density of detected sources matches well with the angular resolution of 12 arcmin (FWHM). There are only few regions where source blending becomes an issue and among them the central square degree of the Galaxy is worth mentioning.

The second catalogue [5] of soft γ-ray sources detected by INTEGRAL lists 209 objects detected with high significance. In the Milky-Way it lists 104 accreting binaries, 4 supernova remnant, 4 pulsars, 2 molecular clouds. The accreting binaries include 78 neutron star systems (31 HMXB, 47 LMXB) plus 2 candidates, 4 black-hole systems plus 10 candidates and 7 cataclysmic variables. The catalogue also contains 55 new sources discovered by INTEGRAL (counterparts/source types have now been proposed for about 20% of them). With improved analysis software currently available and increasing data becoming publicly available one can estimate that INTEGRAL will finally detect not far from 500 hard X-ray sources. For comparison HEAO-1 and SIGMA detected 70 sources down to 14 mCrab and respectively 15 galactic sources to a sensitivity of 30 mCrab. About half of the INTEGRAL detected sources will contribute in the MIRAX FOV centered in the galactic bulge.

The contribution of the point sources detected by INTEGRAL has been compared with the galactic ridge emission measured between 20 and 200 keV. The contribution of the sources accounts for at least 90% of the galactic emission below 40 keV, a fraction decreasing towards higher energies [6]. Above 300 keV the galactic emission is dominated by the positronium emission and above 1 MeV by cosmic ray interactions. Between 20 and 300 keV a hard diffuse component remains that could be accounted for by a population of yet to be detected sources [7].

With the very long exposure that will be avalable in the inner Galaxy and its improved angular resolution, MIRAX will help resolving a larger fraction of the galactic ridge emission into point sources.

THE CENTER OF THE GALAXY

The hard X-ray emission detected close to the galactic center by INTEGRAL [8] has a luminosity of 5×10^{38} erg/s. The peak of the emission is located 1 arcmin from Sgr A*. That emission is much stronger than what could be extrapolated from the soft X-ray emission of Sgr A* and is constant. It very probably corresponds to the hard energy tail of the X-ray diffuse emission observed within 6 arcmin of the galactic center. That hard X-ray emission could be interpreted as synchrotron emission associated with the inverse Compton TeV emission detected by HESS [9]. INTEGRAL also detected the signature for low flux variability at 10 arcmin from Sgr A*, most probably related to an X-ray binary rather than Sgr A*.

Emission above 20 keV has also been detected from the molecular cloud Sgr B2. In that region strong and constant Iron line emission has been detected by ASCA and confirmed by other missions. Together with the X-ray data, the INTEGRAL spectrum of that region show that the emission is dominated by a Compton reflection continuum and fluorescence. The primary source cannot be a transient inside the molecular cloud but could be the signature of activity by Sgr A* some 300 years ago. The luminosity of Sgr A* that could be inferred is 10^{39} erg/s, 10^4 times the present luminosity [10].

Constant monitoring of Sgr A* by MIRAX will hopefully allow to better disentangle the variability from the numerous hard X-ray sources contributing in that region and may allow to detect high luminosity flares that could be expected around the central black-hole of our Galaxy.

TRANSIENT EVENTS

The sky monitoring capabilities of INTEGRAL are providing an average of one Astronomer's Telegram alert issued per week and one Gamma-Ray Burst alert per month.

New source or transient alerts are generated by INTEGRAL once per month on average. At the beginning of the mission this was dominated by bright new persistent (absorbed) sources. Then most alerts concerned transient events, mostly of already known sources. The rate of transient alert is correlated with the periods when INTEGRAL is observing the galactic plane and in particular the galactic buldge. One can estimate therefore that MIRAX will detect more than one transient event per week and will become the main actor monitoring the hard X-ray sky.

Transient alerts from INTEGRAL are triggered manually by the scientist on shift at the INTEGRAL Science Data Center (ISDC – [11]). Automatic alert generation software

are good for very bright sources but, as most transient are detected at a significance below 10 (in short pointings), human screening of the data has proven to be required. A similar set-up will most probably be needed to fully utilize the surveying capabilities of MIRAX.

Among the alerts generated by INTEGRAL one should mention the discovery of many new absorbed HMXB, flares in magnetars, the fastest rotating accretion-powered millisecond pulsar, bright flares of known accreting pulsars and new accreting black-hole candidates. Few very bright (∼1 Crab) events have been observed every year. The possibility to observe continuously those transients with MIRAX with a good time resolution will provide invaluable information on the physics of those sources.

INTEGRAL discovered the fastest rotating accretion-powered millisecond pulsar known to date [12], with a spin period of 1.6 msec. The companion star has an extremely low mass and could be a hot brown dwarf [13], the accretion rate being controlled by gravitational radiation. During the outburst, the source did show spectral evolution that could be understood if the temperature and optical depth of the electrons accelerated in the shock remained almost constant while the size of the hot spot at the surface was decreasing. The increase of the pulse fraction at high energy can be related to Doppler boosting as variability is enhanced where the spectral cutoff occurs [14].

INTEGRAL observed cyclotron lines in several flaring systems with the highest resolution ever. Phase resolved spectroscopy of the cyclotron lines can be achieved on the neutron star spin period timescale. In the case of X0115+63 [15], INTEGRAL detected variations of the line shape and centroid energies for the first time. The continuum slope and cutoff energy vary as well. Such observations allow to map the physical conditions in the neutron star magnetosphere in 3 dimensions. In V0332+53 [16] variation of the pulse profile with energy and between outbursts highlight the stratification of the emission regions in the magnetosphere and their variation over long time scales.

CONTINUOUS LIGHTCURVES OF X-RAY BINARIES

A number of new bright highly absorbed high mass X-ray binaries have been detected by INTEGRAL above 20 keV, doubling the number of supergiant HMXB known in the galaxy. Such sources were either unknown before INTEGRAL or weakly detected in previous X-ray surveys [17].

Super-giant HMXB are strongly (factors >10) variable on time scales of hours. Some are persistent and others are transient [18] (or highly variable and most of the time below the instrument sensitivity). The flux and absorption variability on time-scales of hours to days is related to the dynamics of the stellar wind, to its interaction with the compact object (Fig. 2) and to the orbital inclination. Obtaining continuous lightcurves for a reasonable sample of objects will help classifying and modeling those sources.

Continuous lightcurves of Vela X-1 have been obtained covering two giant flares observed in december 2003 [19]. Those flares are interpreted as dramatic increase in the mass accretion rate related to inhomogeneities in the stellar wind of the companion. The emission modulation and spectra remained however mostly constant indicating that the neutron star magnetosphere was not strongly affected. Quasi-periodic oscillations are suggested during the flares suggesting density structure in the accreted material.

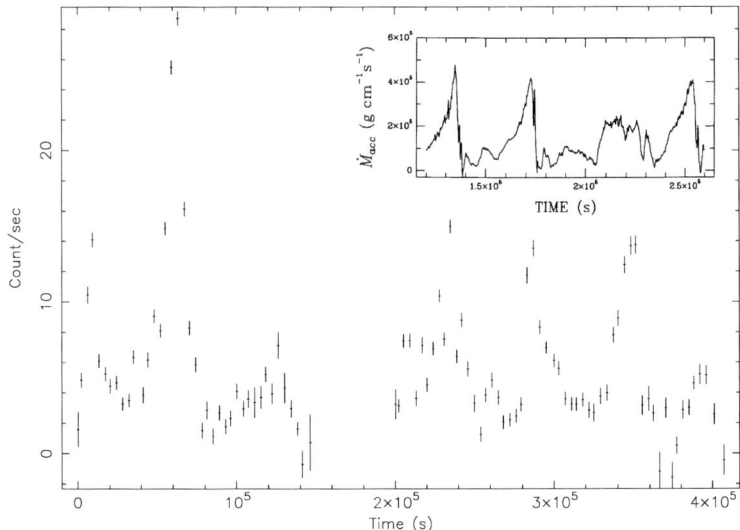

FIGURE 2. Flaring activity observed in IGR J16318-4848 by INTEGRAL (22-60keV ISGRI lightcurve). Inset: mass accretion rate variability driven by accretion wake oscillations (from [20] reproduced by permission of the AAS).

Continuous observations of the inner galaxy by MIRAX should also allow to detect many new fast transient (sg or Be-type) HMXB and to determine long periods easier than with current missions.

The study of LMXB with INTEGRAL is affected by the lack of sensitivity at low energy and by the dithering observation strategy. Continuous lightcurves of bright LMXB obtained with MIRAX will allow to observe all state transitions and their respective duration, something that is poorly observed, and will help understanding accretion in those sources.

REFERENCES

1. P. Ubertini et al., 2003, A&A 411, L131
2. G. Vedrenne et al., 2003, A&A 411, L63
3. N. Lund et al., 2003, A&A 411, L231
4. M. Mas-Hesse et al., 2003, A&A 411, L261
5. A. Bird et al., 2006, ApJ, 636, 765
6. F. Lebrun et al., 2004, Nature, 428, 293
7. A.W. Strong et al., 2005, A&A, 444, 495
8. G. Bellanger et al, 2005, ApJ, 636, 275
9. A. Neronov et al, 2005, astro-ph/0506437
10. M. G. Revnivtsev et al, 2004, A&A 425, L49
11. T. Courvoisier et al., 2003, A&A 411, L53
12. S. Shaw et al., 2005, A&A 432, L13
13. D. K. Galloway et al., 2005, ApJ 622, L45
14. M. Fallanga et al., 2005, A&A, 444, 15
15. A. Santangelo et al., 2005, A&A, submitted
16. I. Kreykenbohm et al., 2005, A&A, 433, L45
17. R. Walter et al., 2005, A&A, submitted
18. I. Negueruela et al., 2005, ESA-SP 604
19. R. Staubert et al., 2004, ESA-SP 552, 259
20. J. Blondin et al., 1990, ApJ, 356, 591

Monitoring Neutron Star High-Mass X-Ray Binaries in the *INTEGRAL* Galactic Plane Survey

J. Wilms

Department of Physics, University of Warwick, Coventry, CV4 7AL, United Kingdom

Abstract. Monitoring neutron star systems allows to measure the flux and pulse period evolution of these systems, which in turn helps in studying their accretion history and the physics of the interaction between the accretion disk and the neutron star's magnetic field. This contribution presents information on the ongoing monitoring activities with the *INTEGRAL* satellite, including an overview of the data presented on the project's web pages and selected scientific highlights.

Keywords: X-ray binaries, neutron stars
PACS: 97.60.Jd, 97.80.Jp, 98.70.Qy

WHY MONITOR NEUTRON STARS?

In accreting neutron star high-mass X-ray binaries, the accreting matter typically forms a small accretion disk. If the neutron star has a high magnetic field, the plasma of the accretion disk will couple at this magnetic field at the Alfvén radius, at a significant distance from the compact object. It then falls along the magnetic field lines onto the poles of the neutron star where it is stopped in a shock front close to the neutron star's surface and subsequently emits in the X-rays and gamma-rays.

X-ray and gamma-ray observations of accreting neutron stars allow us to gain insight into this complex physical environment. Pointed observations, the standard observational strategy chosen by large X-ray and gamma-ray observatories such as *XMM-Newton*, *Chandra*, or *INTEGRAL*, give us a detailed "snapshot" of the physical conditions around a neutron star at a given point in time. Such observations are crucial, e.g., for determining the neutron star magnetic field through through observations of its cyclotron line, or to measure the properties of atomic features below $\sim 10\,\text{keV}$. However, neutron stars are often part of a dynamic environment. For example, the interaction of the neutron star's magnetic field with the surrounding accretion disk will result in changes of the neutron star's spin period on timescales of weeks to years [1]. Furthermore, many neutron star sources are transients, with changes in the mass accretion rate, $\dot{M}$, on long timescales. Finally, especially for wind accretors, the accretion flow is inherently instable, which can result in dramatic changes of the X-ray and gamma-ray properties of these sources on timescales of minutes. To study such dynamic phenomena, monitoring of sources is required, where the same source is repeatedly observed for many years.

With most current satellites, performing such a monitoring is difficult or not possible due to scheduling constraints. It is for this reason that *MIRAX*, a dedicated satellite mission to monitor the Galactic center, is crucially needed. In this contribution, current

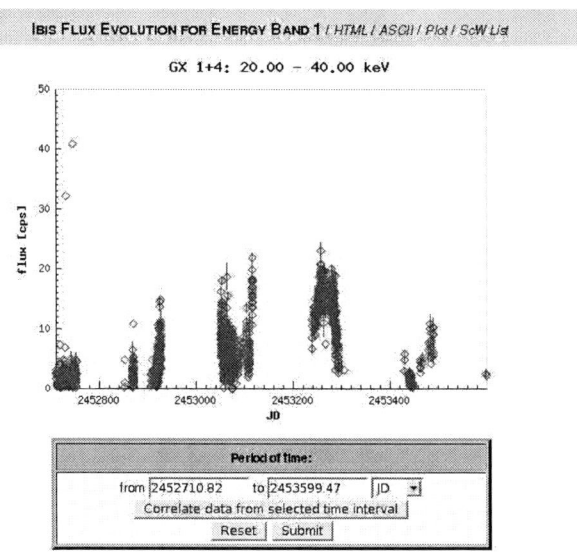

FIGURE 1. Screen shot of the project WWW pages showing the 20–40 keV X-ray light curve of GX 1+4.

efforts to monitor the Galactic center region and the Galactic plane with ESA's *INTEGRAL* satellite are presented, to illustrate what data can be obtained with current day missions. *MIRAX* will greatly outperform these results in the near future.

The remainder of this contribution is structured as follows. First, general information about this monitoring program is given, including information about selected data products from the monitoring which are available online. In the remaining part of this contribution, selected science results from the monitoring are briefly shown.

MONITORING NEUTRON STARS WITH *INTEGRAL*

As part of its guaranteed time program, *INTEGRAL* performs roughly bi-weekly scans of the Galactic plane and a deep survey of the Galactic center in the X-rays and Gamma-rays. In the framework of these observations, an international team composed of the individuals listed in the acknowledgments of this contribution monitors 26 persistent pulsating neutron stars, including sources such as Vela X-1, Cen X-3, or GX 1+4. Fluxes in several energy bands for the production of long-term lightcurves, information about the spectral shape in form of several X-ray colors, information on the detection significance of the source, and other information obtained from these data are presented on the collaboration pages on the World Wide Web (WWW) at http://pulsar.astro.warwick.ac.uk/gps/index.html (mirror sites are in preparation). These WWW pages are updated as new data become available.

Based on these data, the WWW-pages support the on-line generation of light-curves

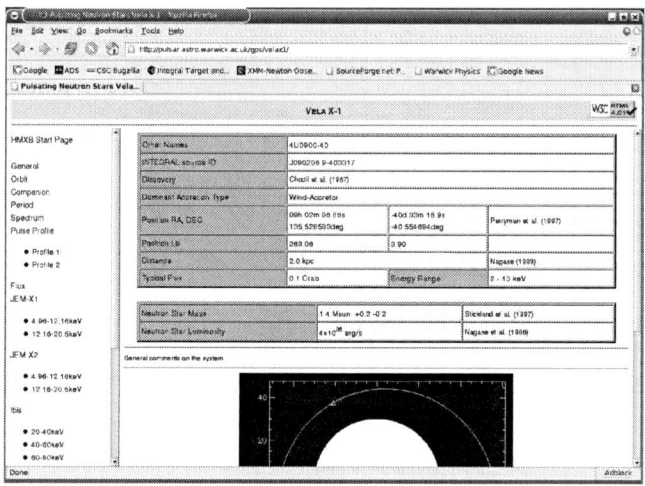

FIGURE 2. Screen shot showing selected information about Vela X-1 from the project WWW pages. At the bottom is the top of a to scale plot of the neutron star's orbit and its companion star.

for *INTEGRAL*'s Jem-X and IBIS instruments in several energy bands on a "science window basis", i.e., based on 1–2 ksec of exposure time. Figure 1 shows a screen shot of the 20–40 keV light curve of the accreting neutron star GX 1+4 (see also Ferrigno et al. [2]). Light curves such as the one shown in Fig. 1 can be generated for user-selected time intervals, specified in the Julian Date, Modified Julian Date, and the *INTEGRAL*-specific *INTEGRAL* Julian Date. Finally, the pages also allow the interactive generation of different color-color and color-flux diagrams.

To help with interpreting the *INTEGRAL* data, the WWW-pages also list background information about each source, including information such as the source position, orbital parameters, the basic properties of the companion, such as its spectral type, luminosity or mass estimates, lists of earlier pulse period measurements, and descriptions of earlier spectral analyses of the source. All of this information is cross-linked to the relevant publications in NASA's Astrophysics Data System. Figure 2 shows an example. Furthermore, the pages also include recent data from the Rossi X-ray Timing Explorer All Sky Monitor (ASM) and will soon also be supplemented by archival data from previous missions such as CGRO BATSE.

SELECTED SCIENTIFIC RESULTS

The *INTEGRAL* monitoring is an ongoing effort[1], however, it has already produced a multitude of results (see, e.g. the contributions by Santangelo, Tomsick, and Walter in

---

[1] From 2006 onwards, monitoring of the plane will cease and the observations will concentrate on the galactic center.

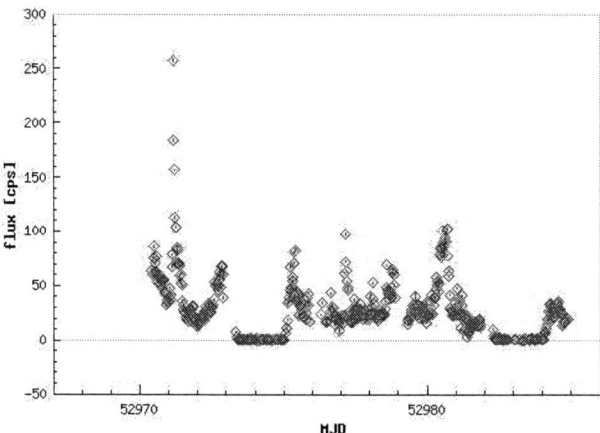

FIGURE 3. Sequence of the December 2003 flaring events of Vela X-1 (screen shot from the project WWW-pages). Phases with count rates of zero are eclipses of the neutron star.

this volume). Here, I will briefly describe two selected highlights: the study of giant flares of Vela X-1 and the period evolution of 4U 1907+09.

Probably due to strong changes in $\dot{M}$, the wind-accretor Vela X-1 shows intermittent strong flaring episodes, where the source can reach fluxes up to $\sim$8 Crab (Fig. 3). These episodes were already seen by *BeppoSAX* (in't Zand, priv. comm.), but it was only with the detection of flares in monitoring observations of the Vela region by *INTEGRAL* [3] that they could be studied in greater detail. During the flares, the X-ray spectrum softens [4, 5], although the pulsed fraction does not change during the flare [6]. Quasi-periodic oscillations with a period of $\sim$2.6 ksec and $\sim$6.7 ksec seem to be present [6]. The rarity of such flares stresses the importance of monitoring sources such as Vela X-1 on long time scales, as will be possible with *MIRAX*.

Finally, in Fig. 4 the *INTEGRAL* derived evolution of the pulse period of 4U 1907+09 is shown. Since at least 1983, this source has been seen to steadily spin down. Recent *RXTE* data suggest that this spin down ceased from the year 2000 onwards [8]. *INTEGRAL* monitoring confirms this result, with no apparent spin up or spin down being measurable to within the uncertainty of the period determination [7]. This result indicates a clear change in the coupling between the accretion disk and the magnetic field of the neutron star, with no apparent change in X-ray spectral shape [8]. It is only with long term monitoring of pulse periods, e.g., with *MIRAX*, that such long-term studies with a much better sampling will be possible in the near future.

ACKNOWLEDGMENTS

I thank the Royal Society for financial support which made attending the conference possible. This contribution is presented on behalf of the INTEGRAL GPS Neutron Star

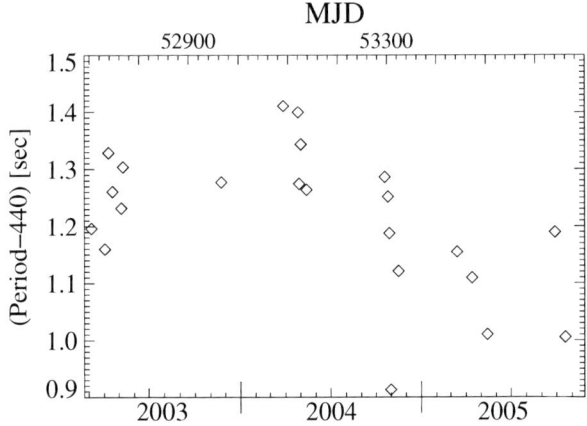

FIGURE 4. Recent period evolution of 4U 1907+09 [7], indicating that the steady spin down of the source has recently stopped and perhaps even changed to a spin up.

Team, consisting of the author and A. Santangelo (IAAT), R. Staubert (IAAT), J. Barnstedt (IAAT), M. Chernyakova (ISDC/Geneva), T. J.-L. Courvoisier (ISDC/Geneva), M. Denis (CBK Warsaw), G. DiCocco (INAF/CNR Bologna), C. Ferrigno (INAF Palermo), S. Fritz (IAAT), P. Kretschmar (ESAC), I. Kreykenbohm (ISDC/IAAT), S. Larsson (Stockholm), J. Maar (IAAT), S. Mereghetti (INAF Milano), A. Paizis (INAF Milano), K. Pottschmidt (UCSD), A. Segreto (INAF Palermo), L. Sidoli (INAF Milano), and N.-J. Westergaard (DSRI Copenhagen).

REFERENCES

1. P. Gosh, and F. K. Lamb, *ApJ* **223**, L83–L87 (1978).
2. C. Ferrigno, A. Segreto, A. Santangelo, J. Wilms, I. Kreykenbohm, M. Denis, and R. Staubert, *A&A* (2006), submitted.
3. R. Krivonos, N. Produit, I. Kreykenbohm, R. Staubert, A. von Kienlin, C. Winkler, and N. Gehrels, ATEL 211 (2003).
4. P. Kretschmar, R. Staubert, I. Kreykenbohm, M. Chernyakova, A. von Kienlin, S. Larsson, K. Pottschmidt, J. Wilms, L. Sidoli, A. Santangelo, A. Segreto, D. Attie, P. Sizun, and S. Schanne, in *ESA SP-552: 5th INTEGRAL Workshop on the INTEGRAL Universe*, 2004, pp. 267–271.
5. I. Kreykenbohm, P. Kretschmar, J. Wilms, K. Pottschmidt, S. Larsson, M. Chernyakova, A.Santangelo, and R. Staubert, *A&A* (2006), in prep.
6. R. Staubert, I. Kreykenbohm, P. Kretschmar, M. Chernyakova, K. Pottschmidt, S. Benlloch-Garcia, J. Wilms, A. Santangelo, A. Segreto, A. v. Kienlin, L. Sidoli, S. Larsson, and N. Westergaard, in *Proc. 5th INTEGRAL Workshop: The INTEGRAL Universe*, ESA SP-552, ESA, Noordwijk, 2004, pp. 259–256.
7. S. Fritz, I. Kreykenbohm, J. Wilms, R. Staubert, F. Bayazit, J. Rodriguez, and A. Santangelo, *A&A* (2006), to be submitted.
8. A. Baykal, S. C. Inam, and E. Beklen, *MNRAS* (2005), submitted (astro-ph/0512176).

RXTE Observations of Galactic Center Transients

C. B. Markwardt

Astroparticle Laboratory, NASA/GSFC Code 661, Greenbelt, MD 20771

Abstract.
We present the results of monitoring observations of the galactic center and plane regions with the *Rossi X-ray Timing Explorer* PCA instrument. Since 1999, we have been performing these monitoring observations twice weekly, and since 2003, the survey solid angle is ~ 500 square degrees. Approximately 150 sources have been detected, of which about 32 were newly discovered. Four millisecond X-ray pulsars have been detected. Sources vary widely in flux as a function of time, and spend a significant portion of their lives in quiescence.

Keywords: galactic bulge RXTE PCA
PACS: 95.85.Nv, 97.60.Gb, 97.60.Jd, 97.60.Lf, 97.80.Jp, 98.35.Jk

INTRODUCTION

Low-mass binaries transfer mass to a compact object at rates depending on their evolutionary state. For many systems we have seen that this rate is in a regime in which the disk accretion is unstable rather than steady, and we learn of the systems through transient outbursts or thermonuclear bursts. Early discoveries were naturally very bright and/or longer lived. Recent RXTE observations have shown that there are an appreciable number of transient sources which are very short-lived, of order 20 days or less. These include four of the seven known accreting millisecond pulsars, but also more exotic sources (and possible black hole systems) such as V4641 Sgr and CI Cam. In order to capture these rare sources as early as possible, we have devised a galactic bulge monitoring program with the *RXTE* Proportional Counter Array.

While we understood that such a new monitoring program would catch new types of sources, we were primarily motivated by the detection of the first millisecond X-ray pulsar, SAX J1808.4−3658 in 1998 [9, 2]. Unlike "typical" X-ray binary outbursts, the millisecond pulsars appear to be intrinsically less luminous ($\sim 5–10\%$ L_{Edd}, a few tens of mCrab for a galactic center source); to have short outburst durations ($\sim$weeks); and long quiescent periods ($\sim$years). The PCA monitoring program was designed to cover a region of the sky with a high density of X-ray binary sources, so as to maximize the chances of detecting more. Quite a number of new sources have indeed been detected. With an eye toward the potential of a *MIRAX* mission, we present results from the RXTE PCA bulge monitoring program.

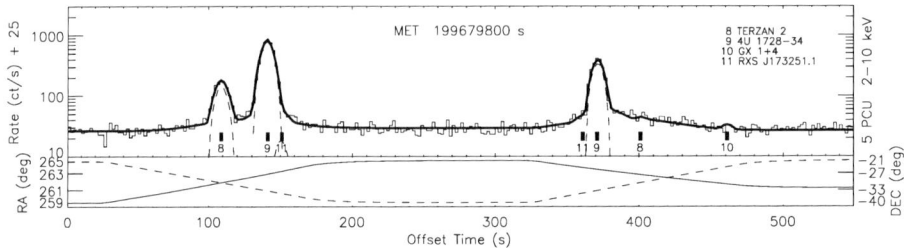

FIGURE 1. Scanning light curve of April 30, 2000. The histogram is the observed light curve, and the thick solid line is the modeled light curve consisting of point sources (numbered marks, dashed lines) and diffuse emission. In the lower panel the spacecraft RA (solid) and DEC (dashed) are shown.

OBSERVATIONS

Monitoring of the galactic bulge region is done with the *RXTE* Proportional Counter Array (PCA) [8]. The PCA instrument has a full effective area of ~ 6500 cm^2, and is sensitive to 2–60 keV X-rays within a collimated field of view, which has a triangular profile and a full width at half-maximum (FWHM) of $1°$. The original scan pattern covered approximately 250 square degrees of the galactic center region twice weekly since February 1999, except for four months when sun constraints interfere. Individual source count rates are modulated by the PCA collimator as they pass into and out of the field of view (Figure 1). The resulting light curves are fitted to a model of known sources and diffuse emission, convolved with the collimator response function. The nominal 1σ sensitivity to variations is approximately 0.5–1 mCrab. While the figure shows a few well-resolved source peaks, it is possible to disentangle many nearby sources as long as they are separated by $> 10'$, since multiple scans cover one position.

Since 2003, an "enhanced" scanning program has been undertaken, which expands the scan region to $\pm 26°$ galactic longitude. In the galactic center region, the scans cover to $\pm 12°$ galactic latitude. Farther from the galactic center, the coverage is $\pm 4°$ latitude. Now covering close to 500 square degrees twice weekly, the PCA bulge monitoring program is a very wide field of view, high sensitivity technique to discover new sources.

RESULTS

Although *RXTE* is not an imaging instrument *per se*, the raster scans of the galactic bulge region can be assembled into an effective map of the region. Such maps have a spatial resolution comparable to the PCA collimator field of view. An example false color map of the extended galactic bulge is shown in Figure 2.

At the present writing, about 150 sources have been detected in the total survey region, of which about 32 were newly discovered (or co-discovered) by the bulge scans. Of the 150 sources, about 45% are neutron star candidates, 15% are black hole candidates, and the remainder are as yet unclassified or diffuse sources.

Here we discuss some individual results briefly.

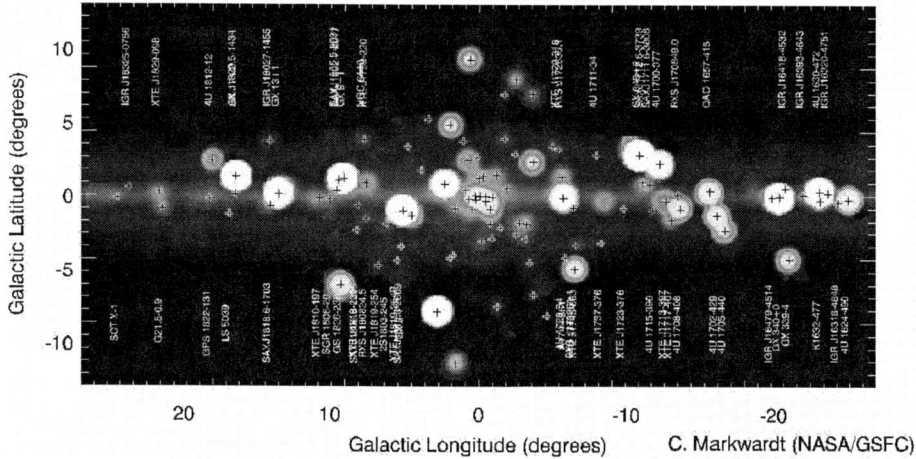

FIGURE 2. RXTE PCA 2–10 keV image of the galactic center and plane region, synthesized from observations on May 24–29, 2003. The original scan path includes the diamond shaped region at the center. The "extended" regions out to longitude 26° started in RXTE Cycle 10.

Milisecond X-ray Pulsars

The bulge scans have been quite successful at detecting millisecond X-ray pulsars. Of the seven known pulsars, three were discovered by the scans (XTE J1751−305, XTE J1814−338 and XTE J1807−294). In addition, several recurrences of the first known pulsar (SAX J1808.4−3658) have been detected. All four were detected confidently by the galactic bulge scans early in their outbursts. Of the four, two are in ultra-compact binaries (period ∼40 min, companion mass ∼0.01–0.02 $M_\odot$) and two are in somewhat larger binary systems (2 and 4 hours). Some of the outbursts have lingered for several months while others have been 2–3 weeks with rapid decays. The other three millisecond pulsars lie outside of the survey region, but would have been detected. The presence of millisecond X-ray pulsars had long been predicted [5, 1], but their detection was elusive. This newly detected population has lead to new understanding of formation and evolution of X-ray binary systems [7].

Eclipsing Low Mass Systems

In contrast to high mass X-ray binary systems, there are few known eclipsing low mass X-ray binary systems. One known system, GRS 1747−312 (in Terzan 6), lies in the galactic center region ($P_{\mathrm{orb}} \sim 12$ hr). Monitoring observations revealed its ∼140 day quasi-regular outburst recurrence pattern, which allowed follow-up observations to be efficiently scheduled. These observations have lead to new constraints on the binary orbital evolution timescale [10]. A second system, XTE J1710−281, was discovered by a scan over the galactic center region ($P_{\mathrm{orb}} \sim 2$ hr). This source appears to be a distant

(> 12 kpc) but highly variable source. Again, bulge monitoring observations have been used to schedule follow-up observations, used to perform eclipse timing. Both sources are also bursters and dippers, making them quite interesting.

Black Hole Outbursts

There have been a number of black hole candidate outbursts in the galactic center region in the past years. Four new black hole candidates were discovered or recently rediscovered after a long period of quiescence (GRO J1655−40, XTE J1720−318, SLX 1746−331 and H1743−322). In some cases, e.g. H1743−322, *INTEGRAL* has discovered the source, and the PCA bulge scans have confirmed the result.

Other Results

- Discovery of eclipses and the binary orbital period of the high mass X-ray binary EXO 1722−361 [3];
- Detection of the X-ray pulsar XTE J1829−098 [6];
- Detection of the anomalous X-ray pular XTE J1810−197 [4].

POPULATIONS OF X-RAY BINARIES

The PCA bulge monitoring program represents a unique sample of the fluxes and variability of sources near the galactic center and galactic plane. While it is customary to construct a $\log N$ vs. $\log S$ type of curve, representing the number of sources (N) detected above a certain cumulative flux limit (S), we have additional information regarding the variability and duty cycle of each source. For each source in the catalog, we have estimated the minimum, median, and maximum flux. Fluxes that fall below the detectability limit of the scans (approx 0.5–1 mCrab in the 2–10 keV band) are excluded. The resulting distribution is shown in Figure 3 (left). The turn-over around 800 mCrab corresponds to a luminosity of $\sim 10^{38}$ erg s^{-1} at the galactic center, i.e. comparable to the Eddington luminosity for a neutron star. As expected based on the known identifications, it is likely that we are seeing primarily neutron star systems, with a handful of black hole systems. A key conclusion to draw from this figure is that X-ray sources in the galactic center region vary by a large amount. The minimum to maximum flux ratio is approximately 3–5. Thus, a short term monitoring program, or infrequent snapshots, are likely to miss many sources near the limiting flux level.

Another way to measure the variability of the bulge sources is to consider their duty cycle. Figure 3 (right) shows the cumulative distribution of duty cycles (where I define the duty cycle to be the fraction of time that the source is detectable). While 20% of the sources detectable all of the time, only 40% of the sources are detectable more than half the time. A significant fraction — 50% — of the sources are detectable one quarter of the time or less! Thus, high on-time monitoring programs will have the greatest success in detecting new sources and new outbursts of previously known sources.

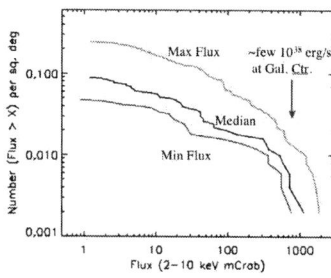 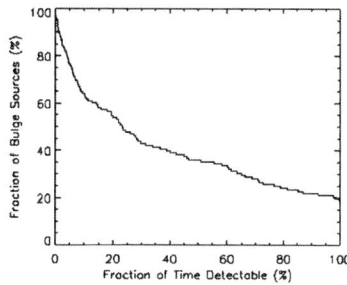

FIGURE 3. RXTE flux and variability distributions for galactic bulge sources. Left: Cumulative distribution of fluxes for sources at the minimum flux (bottom); median flux (middle) and maximum flux (top). The distribution has been converted to per unit solid angle by assuming a constant density in the survey region. Right: Cumulatve fraction of the galactic bulge sources that are detectable (above $\sim$0.5 mCrab 2–10 keV) as a function of the fraction of the time that they are detectable.

PROSPECTS FOR *MIRAX*

The proposed observing strategy for the *MIRAX* mission — long staring observations of the galactic center region with reasonably sensitive instruments — appears to be the right one to detect new transient behaviors. If the success of the PCA bulge scan observations is an indicator, it is likely that there are several new millisecond X-ray pulsars and black hole transients waiting to be discovered.

ACKNOWLEDGMENTS

This work is a product of a large team of co-investigators, including Jean Swank, Frank Marshall, Ron Remillard, David Smith, Tod Strohmayer, Jean in 't Zand; and also the excellent science planning efforts of Evan Smith and Divya Pereira.

REFERENCES

1. Alpar, M. A., Cheng, A. F., Ruderman, M. A., & Shaham, J. 1982, Nature, 300, 728
2. Chakrabarty, D. & Morgan, E. H. 1998, Nature, 394, 346
3. Corbet, R. H. D., Markwardt, C. B., & Swank, J. H. 2005, ApJ, 633, 377
4. Ibrahim, A. I., et al. 2004, ApJ Letters, 609, L21
5. King, A. R. 2000, MNRAS, 315, L33
6. Markwardt, C. B., Swank, J. H. & Smith, E. A. 2004, Astronomers Telegram, 317
7. Podsiadlowski, P., Rappaport, S., & Pfahl, E. D. 2002, ApJ, 565, 1107
8. Swank, J. & Markwardt, C. 2001, in ASP Conf. Ser. 251, New Century of X-ray Astronomy, ed. Inoue, H. & Kunieda, H. (San Francisco: ASP), 94
9. Wijnands, R. & van der Klis, M. 1998, Nature, 394, 344
10. in't Zand, J. J. M., et al. 2003, A&A, 406, 233

Accretion-powered Millisecond Pulsar Outbursts

Duncan K. Galloway

School of Physics, University of Melbourne, VIC 3010 Australia

Abstract. The population of accretion-powered millisecond pulsars (AMSPs) has grown rapidly over the last four years, with the discovery of six new examples to bring the total sample to seven. While the first six discovered are transients active for a few weeks every two or more years, the most recently-discovered source HETE J1900.1−2455, has been active for more than 8 months. We summarise the transient behaviour of the population to estimate long-term time-averaged fluxes, and equate these fluxes to the expected mass transfer rate driven by gravitational radiation in order to constrain the distances. We also estimate an upper limit of 6 kpc to the distance of IGR J00291+5934 based on the non-detection of bursts from this source.

Keywords: X-ray binary, pulsar, accretion
PACS: 97.80.Jp

INTRODUCTION

Each of the known AMSPs have now been well-studied in followup observations, and a review of their observational properties can be found in [1]. Six of the seven are transients with outburst intervals of 2 or more years. However, activity of the most recently-discovered example, HETE J1900.1−2455 [2], has continued long beyond the usual active interval for the other six sources. The latest observation, on 2006 February 2, indicates that the source is still at approximately the same flux level as it was throughout the second half of 2005. The source is also unusual in that the pulsations are not detected consistently while the source is active [3].

Here we present a summary of the outburst history of the accretion-powered MSPs, in order to compare the long-term accretion rates. We also introduce a new method which can give an upper limit on the distance for sources where no bursts have been detected.

OBSERVATIONS AND ANALYSIS

We analysed observations of the AMSPs made with the *Rossi X-ray Timing Explorer* (*RXTE*). We used measurements of the persistent flux and peak flux of thermonuclear (type I) bursts (where available) tabled in the catalog of Galloway et al. (2006a, in preparation). The data were analysed with LHEASOFT version 5.3, released 2003 November 17. The persistent flux was measured by averaging the integrated flux from the best-fit absorbed blackbody plus power-law model in the energy range 2.5–25 keV, to spectra extracted separately for each PCU. We used a bolometric correction to the 2.5–25 keV flux based on absorbed comptt model fits to combined PCA and HEXTE data. The X-ray colors for individual AMSPs were relatively constant over each outburst, and so we adopted a constant correction for each source: XTE J1814−338, 1.86; SAX J1808.4−3658,

2.12; XTE J0929−314, 1.80; XTE J1751−305, 1.66; XTE J1807−294, 1.57; and IGR J00291+5934, 2.54.

RESULTS

We estimated the fluence for each outburst using public *RXTE* PCA and ASM measurements. For intervals where the outburst was not covered by PCA observations, we integrated the ASM intensities instead, using a linear cross-calibration between the PCA and the nearest 1-day average 2–10 keV ASM intensities. XTE J1807−294 and XTE J1751−305 lie towards the Galactic center, and a cataclysmic variable is within the 1° *RXTE* field-of-view centred on IGR J00291+5934. Thus, for those sources we subtracted out a baseline level of 10, 5 and 5×10^{-11} ergs cm^{-2} s^{-1} respectively, which we attribute to contributions from diffuse background and/or unrelated field sources. The fluences derived by this method, scaled to give the estimated bolometric values, are listed in Table 1. We propagated the errors from the uncertainties on individual ASM/PCA measurements. We note that the calculated fluences were generally consistent with prior estimates, to within the uncertainties.

The time-averaged accretion rate driven by angular momentum loss arising from gravitational radiation from the binary is given by [4]

$$\dot{M}_{GR} \gtrsim 3.8 \times 10^{-11} \left(\frac{M_C}{0.1 M_\odot}\right)^2 \left(\frac{M_{NS}}{1.4 M_\odot}\right)^{2/3} \left(\frac{P_{orb}}{2\,\mathrm{hr}}\right)^{-8/3} M_\odot\,\mathrm{yr}^{-1} \quad (1)$$

where M_C is the minimum companion mass, M_{NS} the neutron star mass, and P_{orb} the binary orbital period. Because pulse timing allows measurement only of the projected semimajor axis $a_X \sin i$, only a lower limit on M_C is available. Thus, on equating the time-averaged X-ray flux $\langle F_X \rangle$ and $\dot{M}_{GR}$, we derived lower limits on the distance d for the interval prior to each outburst (Table 1).

For the three sources with thermonuclear bursts, independent estimates of the distance can be made from the peak burst flux. Based on bursts observed by *BeppoSAX*, [5] estimated $d = 2.5$ kpc for SAX J1808.4−3658 or up to 3.3 kpc for a pure He burst. Similarly, [6] estimated $d = 5$ kpc for HETE J1900.1−2455 based on a burst observed with *HETE-II*. While the brightest burst from XTE J1814−338 was not conclusively shown to exhibit radius expansion, the implied $d < 8(10)$ kpc for $X = 0.7(0.0)$ [7]. Based on these distances, or the limits from Table 1 for those sources with no detected bursts, we estimated the long-term averaged $\dot{M}$ for each of the AMSPs (Fig. 1).

We note that the measured outburst fluences and intervals for SAX J1808.4−3658 indicate that $\langle F_X \rangle$ (and hence $\dot{M}$) is decreasing steadily. Since the distance depends on the flux only to the $-1/2$ power, the derived limit varied only by 30%, up to a maximum of 2.9 kpc (Table 1). IGR J00291+5934 is the only other source for which multiple outbursts have been identified, and the increasing outburst interval suggests that $\dot{M}$ may also have been decreasing with time.

Markedly different behaviour has been exhibited by HETE J1900.1−2455 since its discovery in 2005 June [2]. Although the source was too close to the sun for observations during 2005 December and 2006 January, activity has apparently continued for more

TABLE 1. Outburst properties and distance limits for the millisecond X-ray pulsars

Source	Outburst	Start (MJD)	Interval (yr)*	Fluence†	$\langle F_X \rangle$**	Distance limit (kpc)‡
XTE J1807−294	Feb 2003	52681	> 7.1	3.1 ± 0.2	< 1.4	4.7
XTE J1751−305	Jun 1998	50978	> 2.4	...	(< 3.0)	(6.2)
	Apr 2002	52363	3.8	2.3 ± 0.3	1.9	(7.8)
XTE J0929−314	Apr 2002	52376	> 6.3	5.4 ± 0.3	< 2.7	(3.6)
SAX J1808.4−3658	Sep 1996	50333	> 0.67	7.7 ± 0.6	< 36	1.4
	Apr 1998	50911	1.58	5.2 ± 0.5	10	2.5
	Jan 2000	≈ 51547	1.74	5.4 ± 1.7	9.8	2.6
	Oct 2002	52559	2.8	6.2 ± 0.4	7.0	3.1
	June 2005	53522	2.6	4.9 ± 0.6§	5.9	3.4
IGR J00291+5934	Nov 1998	51143	> 2.9	...	(< 1.8)	(4.3)
	Sep 2001	52163	2.8	...	(1.8)	(4.2)
	Dec 2004	53341	3.2	1.63 ± 0.16	1.6	4.5
XTE J1814−338	Jun 2003	52789	> 7.4	2.99 ± 0.12	< 1.3	10.5

* The epoch for the outburst prior to the first known is assumed to be earlier than the first ASM measurements (typically 1996 January 6 or MJD 50088).
† Bolometric fluence, in units of 10^{-3} ergs cm^{-2}
** Estimated time-averaged bolometric flux in units of 10^{-11} ergs cm^{-2} s^{-1}.
‡ The values or limits in parentheses are based on an assumed fluence for the outburst, in the cases where the fluence of only one outburst has been measured with any precision.
§ The fluence for the June 2005 outburst of SAX J1808.4−3658 was estimated from the ASM observations alone, since no public PCA data were available

than 8 months. While the estimated $\dot{M}$ in outburst (based on the approximately constant flux level of $\approx 9 \times 10^{-10}$ ergs cm^{-2} s^{-1} since 2005 June 14 [8]) is just 2% $\dot{M}_{Edd}$ (for $d = 5$ kpc), continuing activity would make this the AMSP with the highest average $\dot{M}$ by far (Fig. 1).

Distance upper limits for non-bursting AMSPs. While thermonuclear bursts have not been detected from four of the AMSPs, we expect that this is because they have been missed in data gaps rather than being absent altogether, as in (e.g.) the high-field pulsars. *RXTE* is in a low-Earth orbit with a period of ≈ 90 min, and suffers regular interruptions when observing most of the sky due to Earth occultations, as well as observations of other sources and passages through regions of high particle density, which introduce additional gaps. The AMSPs be arbitrarily distant, because the implied $\dot{M}$ would exceed the Eddington limit; we may however infer a lower limit, at which point the implied $\dot{M}$ would be high enough to produce sufficiently frequent X-ray bursts that it would be highly improbable that they would all be missed by the *RXTE* observations.

The key factors to determine the likelihood of burst detection are the time density of observations (duty cycle) and the underlying burst rate, which depends in turn on $\dot{M}$ and the H-fraction in the accreted fuel, X_0. Three of the four AMSPs in which no bursts have been detected are in "ultracompact" binaries with $P_{orb} \approx 43$ min. The Roche lobes in such tiny binaries cannot contain a main-sequence companion, indicating that the mass donors are evolved and (probably) H-poor (e.g. [9]). The expected burst recurrence times are thus very long due to the absence of heating from persistent H-burning between

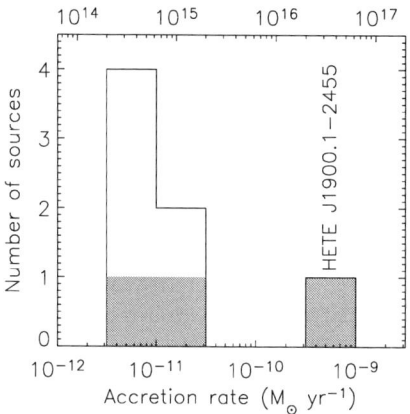
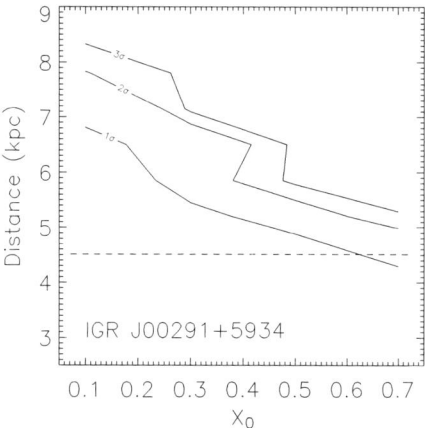

Figure 1: Distribution of time-averaged $\dot{M}$ for the AMSPs. The shaded histogram shows the distribution for sources with distances measured from the peak flux of thermonuclear bursts; the other values are from lower limits on d. The estimated $\dot{M}$ for HETE J1900.1−2455 in outburst, based on the reported source flux and distance, is indicated. The top y-axis is in units of g s^{-1}.

Figure 2: Combined distance-hydrogen fraction limits based on the December 2004 outburst of IGR J00291+5934. The dashed line indicates the lower limit on the distance based on measurements of the time-averaged flux (Table 1. The contours indicate the estimated likelihood that all the bursts were missed for each combination of X_0 and d.

the bursts (e.g. [10]). The combination of very low expected burst rates for these sources, and low duty cycles for the PCA observations (e.g. 6.6% for XTE J1807−294) makes it difficult to constrain the distances. The duty cycle for the PCA observations of IGR J00291+5934, on the other hand, was higher, at 27%. Since the mass donor in this source is also thought to be H-rich, we expect the highest burst rate of all four sources with no detected bursts, and thus is the least likely to have missed all the bursts.

We generated plausible burst sequences for IGR J00291+5934 based on the cubic-spline interpolated flux evolution measured during the 2004 December outburst, with burst ignition conditions calculated as in [10], to which we refer the reader for further details. We adopted a grid of distances beginning at the lower limits in Table 1. We assumed a 1.4 $M_\odot$ neutron star with radius $R - 10$ km, giving a surface gravity $g = (GM/R^2)(1+z) = 2.45 \times 10^{14}$ cm s^{-2} and redshift $1+z = 1.31$. We generated 10^4 burst sequences for each value of d and $X_0 = 0.1, 0.3, 0.5$ and 0.7. We fixed $Z_{CNO} = 0.016$ (equivalent to solar metallicity) throughout. We varied the start time of the first burst evenly within the first predicted burst interval, and also introduced a modest degree of scatter on the burst times, with a standard deviation of 0.13 hr We then checked how many of the predicted burst times fell within the intervals during which the PCA was observing the source. We interpreted the fraction of trials which resulted in one or more detected bursts, as the probability that we could reject that set of parameters.

At the lowest value of $X_0 = 0.1$, the predicted burst rate was sufficiently low that the likelihood of missing any bursts present was high, even for distances as large as 8 kpc. However, for higher values of X_0, the higher $\dot{M}$ implied by such large distances made it increasingly unlikely that we would have missed all the bursts. Thus, the likely distance limit became smaller. Since we expect the mass donor in IGR J00291+5934 is H-rich, we expect a source distance of no more than 6 kpc (Fig. 2).

DISCUSSION

We have derived distances or limits to the known AMSPs via analysis of *RXTE* observations. Based on the outburst history of the AMSPs, the next outburst expected is from XTE J1751−305, early in 2006. In both cases where more than two outbursts are known (SAX J1808.4−3658 and IGR J00291+5934), we find evidence that the long-term averaged flux is decreasing. For the two transients with independent constraints on the distance from the peak flux of photospheric radius-expansion bursts, SAX J1808.4−3658 and XTE J1814−338, the maximum lower distance limit derived from equating the average flux and $\dot{M}_{GR}$ is just above the distance range derived from the bursts. We also derived an upper limit on the distance to IGR J00291+5934 of 6 kpc, based on the predicted burst rate and the duty cycle for the *RXTE* observations. Given a sufficiently high observational duty cycle, this method may be used to derive distance limits on other LMXBs where bursts have not been detected.

Finally, we note the discovery of HETE J1900.1−2455, the first "quasi-persistent" AMSP. As of 2006 February this source is still active, more than 8 months after its discovery. Although the estimated $\dot{M} \approx 2\% \, \dot{M}_{Edd}$ is significantly lower than the peak reached by most of the other transient AMSPs, the fact that activity is continuous indicates that the long-term time-averaged $\dot{M}$ of this source may exceed all the others. In that case, HETE J1900.1−2455 is the best candidate for the detection of gravitational waves from an AMSP.

ACKNOWLEDGMENTS

We thank Lars Bildsten and Philip Podsiadlowski for useful discussions. This research has made use of data obtained through the HEASARC Online Service, provided by NASA/GSFC. This work was supported in part by the NASA LTSA program under grant NAG 5-9184.

REFERENCES

1. R. Wijnands, in *Proceedings of the 2nd BeppoSAX Conference, Amsterdam, 5–9 May 2003*, edited by E. P. J. van den Heuvel et al. 2004, vol. 132, pp. 496–505.
2. R. Vanderspek, et al. *The Astronomer's Telegram* **516** (2005).
3. P. Kaaret, E. Morgan, and R. Vanderspek, *The Astronomer's Telegram* **538** (2005).
4. L. Bildsten, and D. Chakrabarty, *ApJ* **557**, 292–296 (2001).
5. J. J. M. in 't Zand, et al. *A&A* **372**, 916–921 (2001).
6. N. Kawai, M. Suzuki, for the HETE Team, *The Astronomer's Telegram* **534** (2005).
7. T. E. Strohmayer, et al. *ApJL* **596**, L67–L70 (2003).
8. D. K. Galloway, et al. *The Astronomer's Telegram* **657** (2005).
9. P. Podsiadlowski, S. Rappaport, and E. D. Pfahl, *ApJ* **565**, 1107–1133 (2002).
10. A. Cumming, and L. Bildsten, *ApJ* **544**, 453–474 (2000).

What can we learn from long term monitoring of X-ray bursters?

Andrew Cumming

Physics Department, McGill University, 3600 rue University, Montreal QC, H3A 2T8, Canada

Abstract. The last few years have seen the discovery of a number of new aspects of Type I X-ray bursts: the extremely energetic and long duration superbursts, intermediate duration bursts at low luminosities, mHz QPOs, and burst oscillations. These discoveries promise a new understanding of nuclear burning on accreting neutron stars, and offer a chance to use observations to probe neutron star properties. I discuss what we can learn from future long term monitoring with MIRAX.

INTRODUCTION

The fate of matter accreted onto a neutron star is an old question [22, 12], but also an important one. A neutron star in a low mass X-ray binary (LMXB) likely accretes enough mass in its lifetime to replace the entire crust, affecting the long term spin, thermal, and magnetic field evolution of the star. Also, observations of nuclear processing of accreted material offer a chance to probe properties of the neutron star, such as interior thermal structure, magnetic field, and spin. Our understanding of nuclear burning on the surfaces of accreting neutron stars has improved significantly in the last few years. Observations of X-ray bursters with BeppoSAX and the Rossi X-ray Timing Explorer (RXTE) have revealed a host of new phenomena (see [27] for a review), including millisecond oscillations during Type I X-ray bursts, the long duration and energetic "superbursts", long duration X-ray bursts from faint sources, and mHz quasi-periodic oscillations in the persistent emission that are likely due to nuclear burning. In this article, I discuss two aspects which will be significantly impacted by long term monitoring with MIRAX.

PROBING NEUTRON STAR INTERIORS WITH SUPERBURSTS

One of the exciting developments in the last few years has been the discovery of superbursts. These are long duration (several hours), rare (recurrence times ~ 1 year), and energetic (10^{42} ergs) X-ray bursts [16], believed to be due to unstable burning of a thick layer of carbon. The basic idea is that the accreted hydrogen and helium burns within hours to days of arriving on the surface of the star, leaving behind a mixture of heavy elements and a small amount ($\sim 10\%$ by mass) of carbon [26, 25]. This mixture accumulates until the carbon starts to burn, triggering the superburst [8, 28].

Theoretical studies of superbursts initially focused on their potential as probes of nuclear physics. Hydrogen and helium burning proceeds by the rp-process [30, 24], a series of proton captures and beta-decays involving heavy nuclei close to the proton drip line. This process naturally explains the ≈ 100 s extended tails observed in some

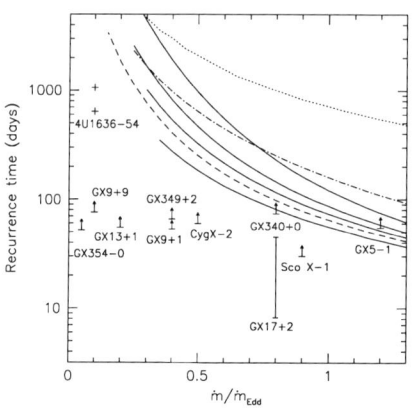 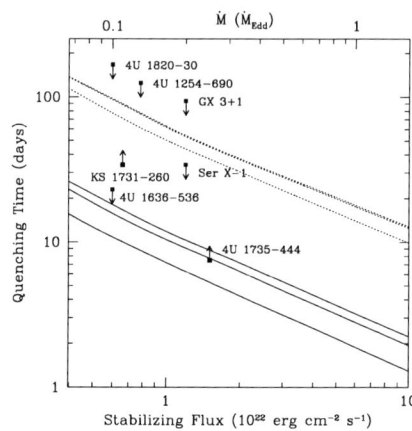

FIGURE 1. Observations of superbursts that will be significantly improved by MIRAX. *Left panel:* Recurrence times compared to theoretical models (from [15]). 4U 1636-54 and GX 17+2 have shown multiple superbursts, allowing a measurement of recurrence time. For the other sources, we show lower limits from BeppoSAX. *Right panel:* Quenching of normal Type I bursts following a superburst (from [9]). We show the predicted quenching time for $y_b = 10^{12}$ (solid lines) and 10^{13} g cm^{-2} (dotted lines) as functions of the critical flux needed to stabilize H/He burning, and the accretion rate. For each value of y_b, the curves are for (bottom to top) energy releases of 1, 2 and 3×10^{17} erg g^{-1}. The data points are taken from Table 1 of [16], with an updated value for 4U 1636-53 (Kuulkers, private communication).

X-ray bursts (e.g. from the regular burster GS 1826-24 [11, 7]). The properties and reaction rates of the unstable proton-rich nuclei involved in the rp-process are not well known experimentally [24]. Because the heavy element composition sets the thermal conductivity of the accumulating carbon layer, Cumming & Bildsten [8] suggested that the superburst properties would directly probe the composition of the rp-process ashes.

Brown [3] and Cooper & Narayan [4] showed that in fact the ignition conditions are much more sensitive to the thermal properties of the crust and core. This is exciting because it gives a new way to probe the neutron star interior, complementary to observations of transiently-accreting neutron stars in quiescence [23, 31, 33], or cooling isolated neutron stars (see [33] for a recent review). They found that to match observed superburst recurrence times required inefficient neutrino cooling in the core, and a low thermal conductivity in the crust.

We recently revisited this question in [10], with two main improvements: (1) using models of superburst lightcurves to constrain the ignition depth, and (2) including neutrino emission from the crust from the formation of Cooper pairs [32]. Cumming & Macbeth [9] modeled superburst lightcurves, finding that the luminosity decays as a broken power law, with break time corresponding to the thermal time at the base of the layer. Fitting these model lightcurves to the observed superbursts constrains the ignition depth to be $\approx 10^{12}$ g cm^{-2} [10]. The temperature required to ignite carbon at this depth is $\approx 6 \times 10^8$ K. Without Cooper pair neutrino emission from the crust, we find that achieving this temperature requires a poor crust conductivity and the inefficient core neutrino emission (modified URCA or smaller), in agreement with previous work. However, surprisingly, including Cooper pair emission makes the crust too cold to achieve ignition at the inferred depth, even for very inefficient core neutrino emission and a low crust conductivity. *It is not possible to match the observed superburst ignition depth when Cooper pair cooling is included.*

The resolution to this puzzle is not yet clear, but will tell us something about the neutron star interior. The Cooper pair neutrino emissivity may be smaller than current calculations suggest, or there may be an extra heating source in the accumulating fuel layer that is not included in current models. An alternative explanation is that these stars are not neutron stars, but rather "strange stars" [1]. Strange stars do not have an inner crust, and so naturally do not have Cooper pair neutrino emission. An evaluation of this scenario [20] shows that ignition at the observed depth is possible for a wide range of parameters for the strange star core.

There are two aspects of superbursts where MIRAX will have a significant impact, providing new observational constraints to help answer these questions. The first is increasing the sample of known superbursts and measuring or improving limits on recurrence times. Figure 1 summarizes the current observational constraints on superburst recurrence times as a function of accretion rate (see [15]). The lower limits of ≈ 50 days are set by the total exposure time accumulated by BeppoSAX, and will improve by an order of magnitude with MIRAX, allowing much better constraints on theoretical models. The second aspect is measuring quenching times of normal Type I bursts following a superburst. Type I bursts disappear for ≈ 1 month following a superburst, as the heat flux from the carbon layer stabilizes the hydrogen and helium burning [8]. Current observations are summarized in Figure 1, but are not very constraining. MIRAX will easily measure quenching timescales, providing an independent measurement of the thickness of the fuel layer and an additional constraint on the theoretical models [9].

MHZ QPOS AND THE TRANSITION TO STABLE BURNING

LMXBs exhibit a range of periodic and quasi-periodic phenomena, ranging in frequency from very low frequency (mHz) noise to kHz quasi-periodic oscillations (QPOs) (see [29] for a review). This variability has mostly been associated with orbiting material in the accretion flow close to the compact object. In the case of a neutron star accretor, an important question is whether any of these phenomena originate from or are associated with the neutron star surface. This is important for identifying the compact object as a neutron star or a black hole, as well as offering a probe of the neutron star surface layers.

Revnivtsev et al. [21] discovered a new class of mHz QPOs in three Atoll sources, 4U 1608-52, 4U 1636-53, and Aql X-1, which they proposed were from a special mode of nuclear burning on the neutron star surface rather than from the accretion flow. These mHz QPOs have frequencies in the range 7–9 mHz (timescales of 1.9–2.4 minutes). In 4U 1608-52, a transient source whose luminosity is observed to change by orders of magnitude, the mHz QPO was only present within a narrow range of luminosity, $L_X \approx 0.5$–1.5×10^{37} erg s^{-1}. This is significant because in many X-ray bursters a transition in burning behavior occurs close to this luminosity, from frequent Type I X-ray bursting at low accretion rates to the disappearance of Type I X-ray bursts at high accretion rates (e.g. [6]). This transition is expected theoretically because at high accretion rates the fuel burns at a higher temperature, reducing the temperature-sensitivity of helium burning and quenching the thin shell instability. However, an outstanding puzzle is that theory predicts a transition accretion rate close to the Eddington rate, almost an order of magnitude larger than observed [2].

Paczynski [19] pointed out that near the transition from instability to stability, oscillations are expected because the eigenvalues of the system are complex (see also [18]). Extending his analysis, we recently showed [13] that the mHz QPOs are indeed naturally explained as being due to marginally stable nuclear burning on the neutron star surface. At the boundary between unstable and stable burning, the temperature dependence of the nuclear heating rate and cooling rate almost cancel. The result is an oscillatory mode of burning, with an oscillation period close to the geometric mean of the thermal and accretion timescales for the burning layer, or ≈ 100 s, matching the observed periods. Numerical simulations with the Kepler code confirm this simple analytic understanding. Interestingly, the observed oscillation period depends sensitively on the surface gravity and the accreted hydrogen fraction, giving a new way to probe these parameters. We are currently working to understand in detail the range of luminosities for which mHz QPOs are observed, and what sets the Q value of the oscillation.

Improved understanding of mHz QPOs in particular, and the global burning behavior in general, from long term monitoring could tell us about the geometry of accretion onto the star. If the mHz QPOs are due to marginally stable nuclear burning, the local accretion rate onto the star must be close to the Eddington rate, even though the global accretion rate inferred from the X-ray luminosity is ten times lower. One possibility is that the accreted material covers only part of the neutron star surface at luminosities $L_X \gtrsim 10^{37}$ erg s^{-1}. This would also explain a number of other puzzling features of Type I X-ray bursts. First, it would account for the disappearance of regular bursting at this luminosity. If some fuel "leaked out" from the stably burning region, it could ignite and cause the occasional X-ray bursts seen at high luminosities. Stable H/He burning at global luminosities well below Eddington would also allow carbon production in large enough quantities to power superbursts [25]. Burst oscillations occur preferentially at higher luminosities [17]. It may be that incomplete spreading of fuel promotes the development of burning inhomogeneities. Finally, Atoll sources undergo a transition from the island to banana state of the color-color diagram close to this luminosity. Perhaps a change in accretion geometry affects the distribution of fuel.

CONCLUSIONS

In summary, long term monitoring of X-ray bursters with BeppoSAX and RXTE has revealed new phenomena that would be difficult to observe otherwise. I have concentrated on superbursts, which occur on long timescales, and mHz QPOs, which occur in a narrow luminosity range. Unfortunately, there is no space in this short article to discuss "intermediate duration" bursts, which are also sensitive to the neutron star interior properties (e.g. [14]), low luminosity bursters (e.g. [5]), burst oscillations, or how burst properties can be used to constrain the composition of the accreted material [7]. MIRAX will dramatically improve our understanding of all of these phenomena, which promises to tell us about (1) the physics of high density matter, (2) neutron star spin and magnetic field evolution, (3) the composition of the donor star and therefore evolution of low mass X-ray binaries, (4) nuclear physics at high temperatures and densities, and (5) the geometry of the accretion flow onto the star.

I would like to thank the conference organizers for their hospitality, and Duncan Galloway, Erik Kuulkers, and Jean in 't Zand for discussions. I acknowledge support from an NSERC Discovery Grant, Le Fonds Québécois de la Recherche sur la Nature et les Technologies, and the Canadian Institute for Advanced Research.

REFERENCES

1. Alcock, C., Farhi, E., & Olinto, A. 1986, ApJ, 310, 261
2. Bildsten, L. 1998, NATO ASIC Proc. 515: The Many Faces of Neutron Stars., 419
3. Brown, E. F. 2004, ApJ, 614, L57
4. Cooper, R. L., & Narayan, R. 2005, ApJ, 629, 422
5. Cornelisse, R., Verbunt, F., in't Zand, J. J. M., Kuulkers, E., & Heise, J. 2002, A& A, 392, 931
6. Cornelisse, R., et al. 2003, A& A, 405, 1033
7. Cumming, A. 2004, Nuclear Physics B Proceedings Supplements, 132, 435
8. Cumming, A., & Bildsten, L. 2001, ApJ, 559, L127
9. Cumming, A., & Macbeth, J. 2004, ApJ, 603, L37
10. Cumming, A., Macbeth, J., in 't Zand, J. J. M., & Page, D. 2005, ApJ, submitted (astro-ph/0508432)
11. Galloway, D. K., Cumming, A., Kuulkers, E., Bildsten, L., Chakrabarty, D., & Rothschild, R. E. 2004, ApJ, 601, 466
12. Hansen, C. J., & van Horn, H. M. 1975, ApJ, 195, 735
13. Heger, A., Cumming, A., & Woosley, S. E. 2006, ApJ, submitted (astro-ph/0511292)
14. in 't Zand, J. J. M., Cumming, A., Verbunt, F., van der Sluys, M. V., & Pols, O. R. 2005, A& A, 441, 675
15. Keek, L., in 't Zand, J. J. M., & Cumming, A. 2006, A& A, submitted
16. Kuulkers, E., 2003, in "The Restless High-Energy Universe" (Amsterdam, May 5-8, 2003), ed. E.P.J. van den Heuvel, J.J.M. in 't Zand, and R.A.M.J. Wijers (astro-ph/0310402)
17. Muno, M. P., Fox, D. W., Morgan, E. H., & Bildsten, L. 2000, ApJ, 542, 1016
18. Narayan, R., & Heyl, J. S. 2003, ApJ, 599, 419
19. Paczynski, B. 1983, ApJ, 264, 282
20. Page, D., & Cumming, A. 2005, ApJ, 635, L157
21. Revnivtsev, M., Churazov, E., Gilfanov, M., & Sunyaev, R. 2001, A& A, 372, 138
22. Rosenbluth, M. N., Ruderman, M., Dyson, F., Bahcall, J. N., Shaham, J., & Ostriker, J. 1973, ApJ, 184, 907
23. Rutledge, R. E., Bildsten, L., Brown, E. F., Pavlov, G. G., Zavlin, V. E., & Ushomirsky, G. 2002, ApJ, 580, 413
24. Schatz, H., et al. 1998, Phys. Rep., 294, 167
25. Schatz, H., Bildsten, L., & Cumming, A. 2003a, ApJ, 583, L87
26. Schatz, H., Bildsten, L., Cumming, A., & Wiescher, M. 1999, ApJ, 524, 1014
27. Strohmayer, T. E., & Bildsten, L. 2003, in Compact Stellar X-Ray Sources, eds. W.H.G. Lewin and M. van der Klis (Cambridge: Cambridge University Press) (astro-ph/0301544)
28. Strohmayer, T. E., & Brown, E. F., 2002, ApJ, 566, 1045
29. van der Klis, M. 2004, in "Compact Stellar X-Ray Sources", eds. W.H.G. Lewin and M. van der Klis (Cambridge: Cambridge University Press) (astro-ph/0410551)
30. Wallace, R. K. & Woosley, S. E. 1981, ApJS, 45, 389
31. Wijnands, R., Guainazzi, M., van der Klis, M., & Méndez, M. 2002, ApJ, 573, L45
32. Yakovlev, D. G., Kaminker, A. D., & Levenfish, K. P. 1999, A& A, 343, 650
33. Yakovlev, D. G., & Pethick, C. J. 2004, ARA&A, 42, 169

High Mass X-ray Binaries Pulsars- a brief review at hard X-rays

A. Santangelo

Institut fuer Astronomie und Astrophysik, SAND 1, 72076 Tuebingen, Germany

Abstract. In this paper we briefly review the status of spectral studies of X-ray binary Pulsars, focusing on their complex phase dependent continua and on the role of electron resonant cyclotron scattering features. We present an overview of the observational results and very briefly discuss the current ideas in theoretical efforts.

Keywords: X-rax Binaries; Accreting Pulsars; Cyclotron lines
PACS: 95.30.Jx, 97.80.Jp, 98.70.Qy, 97.60.Gb

X-RAY BINARY PULSARS

Radiation from X–ray Binary Pulsars (XRBPs) originates from the accretion of ionized gas into the plasma atmospheres of strongly magnetized ($B \sim 10^{12}$ Gauss) rotating neutron stars (NSs). The plasma from a primary companion, is threaded at the Alfvén radius [1], along the B-field lines and then funneled in accretion structures, a solid column or a hollow cylinder [3], onto the magnetic poles at the neutron star surface [36], [10]. If the magnetic and rotation axes are disaligned, pulsed emission is observed. XRBPs belong, depending on the mass of the primary star, to Low Mass (LM) or High Mass (HM) X-ray Binary systems (XRBs). In old ($10^7 - 10^9$ yr) LMXRBs the primary is typically a later than type A star ($M \leq M_\odot$) or a white dwarf, while in younger ($10^5 - 10^7$ yr) HMXRBs it is an O or B star with $M \geq 5M_\odot$. HMXRBs have been further divided in two subgroups, those where the primary is a Be star and those where the primary is a supergiant (SG). In the case of supergiant companion the compact star usually accretes, at a rate of $10^{-6} - 10^{-10} M_\odot yr^{-1}$, from a highly supersonic wind with $v_t \sim 1000$ km s^{-1}. However, in brighter SG systems, accretion can be also powered by Roche-lobe overflow via an accretion disk. Be X-ray binaries have relatively wide orbits with moderate eccentricity and therefore the compact companion spends most of its time away from the disk surrounding the Be Star. X-ray outbursts are expected at the periastron passage, in the low velocity and high density wind around the Be star. Since the historic discovery of Cen X–3 [15] $\sim$ 110 HMXRB have been observed in the Galaxy, including almost 60 pulsars, most identified as Be/X-ray binaries [26]. More recently 92 and 36 HMXRBs have been identified in the SMC and LMC respectively, including 47 and 7 pulsars [27].

---

[1] At the Alfvén radius, also called the magnetospheric radius, the ram pressure of a spherically symmetric inflow equals the magnetic pressure. It is given by $r_A = 2.9 \times 10_8 M_1^{1/7} R_6^{-2/7} L_{37}^{-2/7} \mu_{30}^{4/7}$ cm, that is several hundred neutron star radii.

THE HARD SPECTRA OF XRBPS

The continuum

BeppoSAX and *RXTE* observations have shown that broad band 0.1-200 keV spectra of XRBPs are very complex and may contain many different components: 1) a 0.1 keV blackbody soft excess at low energy, due to reprocessing; 2) a $\alpha \sim 1$ power law continuum up to a cut-off energy at $\sim 15-20$ keV; 3) an exponential roll–off above the cut-off energy; 4) emission lines such as the line for cold iron at $\sim$6.4 keV; 5) absorption feature(s), due to cyclotron resonant scattering, at hard energies (see next paragraph); 6) in some cases (X-Persei, [11]; XTE 1946+274, [37]) a second continuum component at higher energies, possibly due to a non-thermal, partially comptonised cyclotron emission (or bremsstrahlung emission) has been observed. Thanks to their high statistics at hard energies, *BeppoSAX* and *RXTE* have also shown that phase-averaged spectra can be considered only a rough approximation of the spectral behavior of XRBPs, that is naturally phase-dependent. Spectra of X–ray pulsars have been observed to strongly vary with the pulse phase (such as in Cen X–3, [6]; X0115+063, [38]; Vela X-1, [24], [21]; GX 301-2, [25], [22]) and different continua are necessary to model the emission from different pulse phase, i.e. from different physical and geometrical region of the pulsar.

Several *phenomenological models* have been developed to describe the continuum of XRBPs. In 1983 Swank & Holt [44] first proposed the "standard" X-ray continuum for XRBPs as a power law with a high energy exponential roll-off above $\sim 5-20$ keV, namely $f(E) = AE^{-\alpha} \times \exp(-E/E_{fold})$, where $f(E)$ is the photon flux, E_{fold} is the e-folding energy and α is the photon index. Based on (4–30 keV) GINGA spectra of Accreting Pulsars, Mihara et al. [32], [29], proposed the Negative and Positive Power Law Exponential (NPEX) model as the standard model for XRBPs: $f(E) = (AE^{-\alpha_1} + BE^{+\alpha_2}) \times \exp(-E/kT)$, which assuming $\alpha_2 = 2$, mimics the saturated inverse Compton spectrum. A power law with Fermi-dirac cut-off has been also extensively used in literature: $f(E) = (AE^{-\alpha} \times 1/(1 + \exp((E - E_{cut})/E_{fold}))$. To have a more physical description of the phase resolved continuum spectrum observed by *BeppoSAX*, a new class of models (PWCOMPT) has been developed so to provide a simple analytical approximation for the spectrum produced by the upscattering of soft photons in a high temperature, T, but optically thin plasma [41]. This is described by $f_1(E) \propto E^{-\alpha}$ if $E < E_c$ and $f_1(E) \propto E^2 \exp(-E/kT_e)$ if $E > E_c$ (smoothed at E_c).

Despite almost three decades of studies *no physical model* has been found to explain the wide band spectra of XRBPs. This reflects the complexity of radiative transport in the highly magnetized plasma of the accretion environment, which requires a full magnetohydrodynamical approach. All the necessary "ingredients" of the physical scenario at the base of the production and emission of of pulsed X-ray from XRBP have been recently discussed by Orlandini, 2004 [34]. Many attempts have been made to numerically simulate XRBP spectra with some qualitatively agreement with the observations [20], [30],[31]. More recently Becker & Wolff (2005a,b, [4], [5]) succeeded in reproducing XRBP broad band continua by introducing the effect of bulk or dynamical Comptonization of and *ad hoc* source of soft photons, typically associated with the emission from a "thermal mound" at the base of the accretion column.

Cyclotron lines

Cyclotron lines or, more precisely, Cyclotron Resonant Scattering Features (CRSFs) provide a powerful tool for directly measuring the high ($B \geq 10^{12}$ G) magnetic field strengths of accreting neutron stars in XRBPs. In such intense fields, electrons move helicodally along the B field lines with the motion of the electron perpendicularly to the B field quantized in the Landau levels: $E_n = (n + 1/2 + s) E_{cyc}$ with n the principal quantum number, s the electron spin and $E_{cyc} = \frac{eB\hbar}{m_e} = 11.6 B_{12} \cdot (1+z)^{-1}$ keV, where B_{12} is the magnetic field strength in unit of 10^{12} G, and z, is the gravitational red-shift in the scattering region given by $(1+z)^{-1} = \left(1 - \frac{2GM_{NS}}{R_{NS}c^2}\right)^{1/2}$. For typical values of M_{NS} and R_{NS}, CRSFs are expected to be observed at hard X–ray energies. At higher magnetic fields, when relativistic effects are taken into account, $E_{cyc} \propto [(1+2n\frac{B}{B_{crit}}sin^2\theta)^{1/2} - 1]/sin^2\theta$ where θ is the viewing angle with respect to B and $B_{crit} = m^2c^3/e\hbar = 4.4 \times 10^{13} G$ is the critical field. This implies a non harmonic line spacing, though with shifts of $\leq 10\%$. CRSFs were first discovered at hard X-ray energies in the spectrum of Her X–1 [42] and, subsequently, in the hard X–ray transient pulsar X0115+63 [43]. Since then CRSFs were detected in other systems with GINGA, that performed the first systematic (4-30 keV) spectral study of XRBPs, HEXE/TTM on Mir [28], OSSE onboard CGRO [16]. CRSFs studies received a new impulse with *BeppoSAX* and *RXTE*, which have observed CRSFs in at least 15 systems, making cyclotron lines a common feature in the hard spectra of XRBP. A good review of the *RXTE* findings can be found in [7] and [19]. As far as *BeppoSAX* is concerned, we refer the reader to [34] and [39]. Before *BeppoSAX* and *RXTE*, relatively little was known on higher cyclotron harmonics. Besides the pioneering detection of two lines in the spectrum of X0115+63 [44] and the controversial detection in A0535+26 [28], the presence of two cyclotron lines has been reported, based on *BeppoSAX* and *RXTE* data, for Vela X–1 [21, 24], 4U1907+09 [8] and marginally in the phase resolved spectra of Cen X–3 and Her X–1 [39]

The most relevant result in the field has been, however, the discovery of phase dependent multiple harmonics, up to five, centered at energies of ~ 12.7 keV, $\sim 24.$ keV, ~ 36 keV, ~ 50 keV, and ~ 60 keV, in the spectrum of the High Mass X–ray binary X0115+063, one of the best studied X–ray pulsating transients [38, 18]. More recently, Pottschmidt et al., [35], have reported the observation of three CRSFs at ~ 27 keV, ~ 51 keV and ~ 74 keV in the phase averaged data of the 2004-2005 outburst of the HMXRB V0332+53. These results strongly provide a very clear confirmation of the magnetic nature of the absorption lines observed in the spectra of accretion powered pulsars. In some cases (X0115+63, Her X–1) the line centroids, appears not equally spaced. Although relativistic corrections can introduce a slight anharmonicity in the Landau levels [17], a closer look at the data reveals that this result can be ascribed to the value of the centroid of the fundamental. First of all, the fundamental lies close to the energy interval where the the slope of the X-ray spectrum steepens rapidly, and therefore the determination of its centroid energy could be affected by a somewhat inadequate modelling of the continuum. Moreover Araya–Gochez & Harding [2, 1], studying the impact of radiation anisotropy on the shape of the lines in the emergent spectra from relativistic

Compton scattering in the high–field regime, have shown how the shape of the wings of the fundamental CRSF strongly depends on the physical parameters of the emission region and viewing geometry and, is actually a key indicator of isotropy/anisotropy of the continuum radiation. Very recently, Schönerr and collaborators [40], have modeled CRSFs using an improved version of the Monte carlo code of [2, 1] including relativistic scattering cross sections and photon spawning. Their preliminary results suggest that observed CRSFs profiles can indeed be reproduced by the code allowing the determination of the main phyisical parameter of the system, as the plasma geometry, the electron parallel temperature, and therefore proving insights into the physical condition of the accretion column.

RESULTS FORM INTEGRAL

While a catalogue of 34 XRBPs observed by INTEGRAL over 2 years of observations have been presented by Filippova et al. [14], the most interesting findings of INTEGRAL include: 1.) **V0332+53**- following the *RXTE* discovery, the observation of multiple lines harmonically spaced, and not strongly phase dependent in the spectrum of the source of 2005 outburst [23]. Moreover, monitoring the decay of the outburst INTEGRAL has also observed the increase of the fundamental line centroid and depth with the decreasing of the luminosity [33]. This suggests either a change in the extent of the CRSF forming region or a variation of the accretion column height with the accretion rate, that according to [3] is given by $H \propto \dot{M}$; 2.) The observation in the spectrum of **A0535+26** of a broad, weak fundamental line at $E_{cyc} \sim 45$ keV and its first harmonic at $E_{cyc}^2 \sim 100$ keV. This proved the pulsar's field to be $B \sim 4 \times 10^{12}$ G instead of the $\sim 10^{13}$ G often claimed in the literature; **X0115+63** The observation of phase dependent multiple harmonics in the broad band spectrum of the source which confirmed previous detections while extending the continuum up to a few hundred keV [12]. **GX1+4** The observation of a phase dependent continuum with marginal evidence for an aborption line at $E_{cyc} \sim 34$ keV in the descending part of the pulse profile of the source together with the observation of spectral variations with luminosity [13].

CONCLUSIONS

Observational studies of the spectral-time behavior of X-ray Binary Pulsars require:
- Broad-band spectra from a few up to hundred keV, heavily weighted at hard energies to disentangle the many components of the spectra;
- High Statistics, to obtain highly significant phase resolved spectra;
- Long term monitoring of the bursting activity in order to study the dependence of the spectral and timing parameters over up to three order of magnitudes in luminosity.

All these requirements appears, indeed, to be satisfied by a mission like *MIRAX*.

REFERENCES

1. R. A. Araya-Gochez & K. A. Harding, ApJ, 517, 334 (1999)
2. R. A. Araya-Gochez & K. A. Harding, ApJ, 544, 1067 (2000)
3. M. M. Basko & R. A. Sunyaev, MNRAS, 175, 395
4. P. Becker & M. Wolff, ApJ, 621, L45 (2005a)
5. P. Becker & M. Wolff, ApJ, 630, 465 (2005b)
6. L. Burderi, et al., ApJ, 530, 429 (2000)
7. W. Coburn et al., ApJ 580, 394, (2002)
8. G. Cusumano, et al., A&A, 338, L79 (1998)
9. D. Dal Fiume, et al., AdSpRes. 25, 399 (2000)
10. K. Davidson and J.P. Ostriker, ApJ, 179, 585, (1973).
11. T. Di Salvo, et al., ApJ, 509, 897 (1998).
12. C. Ferrigno et al., *"INTEGRAL observation of the 2004 Outburst of X0115+63"*, manuscript in prep.
13. C. Ferrigno et al., *"INTEGRAL observation of the accreting pulsar GX 1+4"*, subm. to ApJ
14. E. V. Filippova et al., Astronomy Letters, 30, 824, (2004), astro-ph/0509525
15. R. Giacconi et al., ApJ 167, L67 (1971)
16. J. E. Grove, et al., ApJ, 438, L25 (1995)
17. A. K. Harding & J. K. Daugherty, ApJ, 374, 687 (1991)
18. W. Heindl, et al., ApJ, 521, L49 (1999)
19. W. Heindl, et al., In *X-ray Timing 2003: Rossi and Beyond*, ed. P. Kaaret, F. K. Lamb, and J. Swank (AIP), astro-ph/0403197
20. M. Isenberg, M., D. Q. Lamb, J. C. L. Wang, ApJ, 505, 688 (1998)
21. I. Kreykenbohm, et al., A&A, 395, 129 (2002)
22. I. Kreykenbohm, et al., A&A, 427, 975 (2004)
23. I. Kreykenbohm, et al., A&A, 433, L45 (2005)
24. A. La Barbera, et al., A&A 400, 993 (2003)
25. A. La Barbera, et al., A&A 438, 617, (2005)
26. Q. Z. Liu, et al., et al., A&A Suppl. Ser. 147, 25, (2000)
27. Q. Z. Liu, et al., et al., A&A 442, 1135, (2005)
28. E. Kendziorra, et al., A&A, 291, L31 (1994)
29. K. Makishima et al., ApJ, 525, 978 (1999)
30. P. Mészáros & W. Nagel, ApJ, 298, 147 (1985a)
31. P. Mészáros & W. Nagel, ApJ, 299, 138 (1985b)
32. T. Mihara, PHD Thesis, University of Tokyo,(1995)
33. N. Mowlavi et al., A&A, , Accepted for publication (2006)
34. M. Orlandini, 2004, In *X-ray and Gamm-ray Atrophysics of Galactic Sources* Proc. of Fourth AG-ILE Workshop. ed. M. Tavani, A. Pellizzoni, S. Vercellone, 119, Aracne Editrice (see also astro-ph0510267, 2005)
35. K. Pottshmidt, et al., ApJ634, L97, 2005
36. J. E. Pringle and M. Rees, Astr. Ap., 21, 1 (1972)
37. A. Santangelo, et al., *"BeppoSAX observation of a phase dependent cyclotron line in XTE J1946+274"* subm. to ApJ
38. A. Santangelo, et al., ApJL, 523, L85 (1999).
39. T. Di Salvo, A. Santangelo, A. Segreto, Nucl. Phys. B Proc. Suppl., 132, 446
40. G. Schönerr et al., in Proc. of *The X-ray Universe 2005* El Escorial, Madrid (2005)
41. R. A. Sunyaev & L. G. Titarchuk, A&A, 86 121, (1980)
42. J. Truemper, et al., ApJ, 219, L105 (1978)
43. Wm. Wheaton, et al., Nature, 282, 240 (1978)
44. N. E. White, J. H. Swank & S. S.Holt, ApJ, 270, 711, (1983)

Long-term developments in Her X-1: Correlation between the histories of the 35 day turn-on cycle and the 1.24 sec pulse period

R. Staubert*, S. Schandl*, D. Klochkov*, J Wilms†, K. Postnov** and N. Shakura**

*Institut für Astronomie und Astrophysik – Astronomie, University of Tübingen, Germany
†Department of Physics, Univ. of Warwick, UK
**Sternberg Astronomical Institute, Lomonossov University, Moscow, Russia

Abstract.
   We have studied the long-term (1971–2005) behaviour of the 1.24 sec pulse period and the 35 day precession period of Her X-1 and show that both periods vary in a highly correlated way (see also Staubert et al. 1997 and 2000). When the spin-up rate decreases, the 35 day turn-on period shortens. This correlation is most evident on long time scales ($\sim$2000 days), e.g., around four extended spin-down episodes, but also on shorter time scales (a few 100 days) on which quasi-periodic variations are apparent. We argue that the likely common cause is variations of the mass accretion rate onto the neutron star. The data since 1991 allow a continuous sampling and indicate a lag between the turn-on behaviour and the spin behaviour, in the sense that changes are first seen in the spin, about one cycle later in the turn-on. Both the coronal wind model (Schandl & Meyer 1994) as well as the stream-disk model (Shakura et al. 1999) predict this kind of behaviour.

Keywords: binaries - accreting, X-rays, neutron stars, accretion, Her X-1
PACS: 95.85.Nv, 95.55.Ka

INTRODUCTION

The LMXB Her X-1/HZ Her shows periodicities on very different time scales. The spin period of the neutron star is 1.24 sec, the orbit lasts 1.7 days and the 35 day cycle corresponds to the precession of the warped accretion disk in the tidal field of the companion. The binary system shows a high inclination of more than 80 deg, and therefore the disk covers the neutron star for the observer temporarily during the 35 day precession, resulting in strong variations of the X-ray signal. The underlying clock is not very accurate (Staubert et al. 1983, Klochkov et al. 2006), but we connect its temporal behaviour with changes in the rate of transfer of mass and (more importantly) of angular momentum. This is supported by a strong correlation between the duration of the precession cycle and changes in the spin-up rate where the latter is due to variations of the angular momentum transfer rate (Ghosh & Lamb 1979).

   We suggest a physical explanation within the framework of the coronal wind model (Schandl & Meyer 1994) and/or the stream-disk model by Shakura et al. 1999). A small variation in the mass transfer rate from the companion may be sufficient to alter the torque on the NS and the shape and tilt of the warped disk, changing the spin- and precession-frequency, respectively.

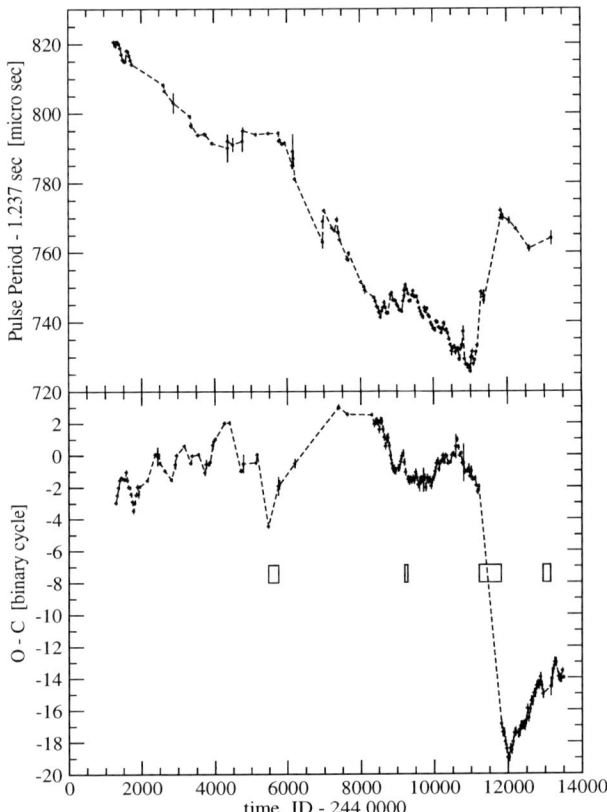

FIGURE 1. The 34 year history of the Her X-1 pulse period (upper panel) and the turn-on data (lower panel), where (O−C) is the difference between observed and calculated turn-on times. The calculated time follows the epoch of Staubert et al. (1983), the 31st turn-on being on T_{31} = JD 2442410.349 d with a 35 day model period of 20.5 P_{orbit} with P_{orbit} = 1.700167788 d (Deeter et al. 1981). – The rectangles mark the periods of the four historical Anomalous Low periods (Parmar et al. 1985, Vrtilek et al. 1994, Parmar et al. 1999, Coburn et al. 2000, Boyd et al. 2004, Still & Boyd 2004).

DATA BASE

The Spin Period Data

The historical data for the 1.24 s X-ray period (the spin of the NS) are compiled from the original literature. A complete list will appear in Staubert et al. (2006). Since 1991 pulse periods were regularly measured by BATSE onboard of CGRO. We have made use of the publically distributed pulsar data as well as lists kindly provided by R. Wilson, and data from our own pointed RXTE observations (e.g. Stelzer et al. 1997, Kuster et al. 2005) and the public RXTE archive. The pulse period development from its discovery until today is shown in Fig. 1 (top). The average spin- up trend dP/dt before day 1150 amounts to $\sim$9 ns/day, but with clear deviations, including episodes of spin-down. A dramatic spin-down event happened around day 1150, associated with the third historical anomalous low period (AL 3).

The Turn-on Data

Her X-1 shows a 35 day flux modulation supposedly due to the occultation of the NS by a precessing warped accretion disk. Turn-ons, the rise of the X-ray signal towards the Main-On, are observed since the discovery of Her X-1 in 1972 (Tananbaum et al. 1972). Our historical data set is based on Staubert et al. (1983). Additionally, we determined turn-ons from the occultation and pulsed flux data of BATSE and from the RXTE All Sky Monitor. These data were taken from the HEASARC archive at NASA/GSFC. Details of the determination of the turn-on times and a complete list of turn-ons will be given in Staubert et al. (2006). The fluctuations of the turn-on times can be expressed by the "(O−C)"-diagram (Fig. 1, lower panel), which shows the difference between the observed turn-on time and the calculated turn-on time assuming a model 35 day period equal to 20.5 P_{orbit} (positive values correspond to turn-ons which are observed later than the calculated one).

OBSERVATIONAL RESULTS

The mean general spin-up of the neutron star ($\sim$9 nsec/d) is modified by significant structure: most apparent are four periods of extended spin-down. These events happen over a time scale of about 2000 days and occur near the times of extended periods of low flux (so called Anomalous Lows, AL), as marked by the rectangles in Fig. 1. In addition, there are deviations on shorter time scales (a few hundred days) which are of quasi-periodic nature (on time scales of 400–600 days, see Fig. 3).

The (O−C)-diagram is a representation of the history of the 35 day period. It appears from Fig. 1 that on the longest time scale the average period is consistent with 20.5 times the binary period, a prediction made 22 years ago by Staubert et al. (1983) when only the first increasing leg in the (O−C)-diagram was known. Individual 35 day cycles (from one turn-on to the next) are either 20, 20.5 or 21 times the binary period, with a few cycles with 19.5 times the binary period just before the onset of anomalous lows and probably several inside the dramatic AL 3 (see below).

Figs. 1 and 2 show that the large features in the development of the pulse period and in the (O−C) diagram are highly correlated. The following global features appear simultaneously: (1) *maxima in pulse period residuals (PPR)*, corresponding to (relative) spin-down, (2) *minima in (O−C)* (going into the minimum means a short turn-on period), (3) the *appearance of Anomalous Lows* around these extrema.

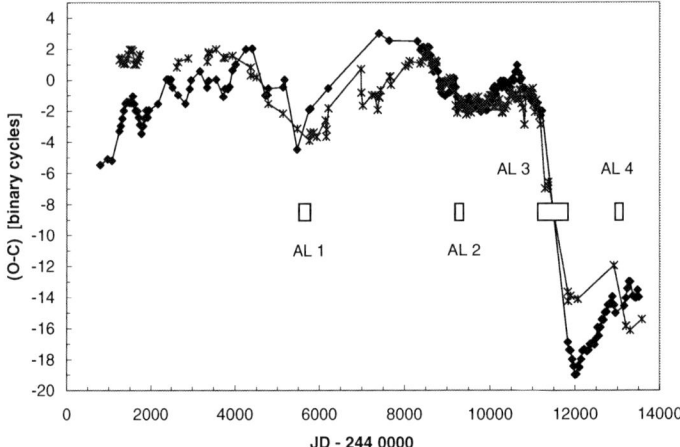

FIGURE 2. The turn-on data (O−C) (diamonds) are repeated from Fig. 1. The second data set (stars) are pulse period residuals (PPR): that is the separation of the measured pulse periods (upper panel in Fig. 1) from a linear function representing a constant spin-up. These residuals were inverted, shifted and scaled in such a way that there is an optimum fit to the (O−C) values for the time frame JD-2440000 between 8500 and 10000. A clear correlation for the global shape of the two curves is evident. The boxes mark the positions and durations of the four anomalous lows (AL).

These global features repeat about every 5 years, with large differences in relative strength and duration and with quite some jitter in the timing. Extrema in (O−C) mean that the average 35 day period changes its value (it increases coming out of a minimum, and decreases after a maximum). The most dramatic event is AL 3 in 1999 (around JD-2440000=1150). The most likely average period during AL 3 was about 33.4 days. Since the start of the observations in 1971, some anomalous lows have apparently been missed because of poor sampling and/or intrinsic weakness or shortness.

There are, however, also *clear correlations on shorter time scales* (a few hundred days), as shown in Fig. 3. We note that this kind of correlation has been seen earlier by Bochkarev et al. (1981) and by Ögelman et al. (1985) on the basis of smaller sets of data. The data in Fig. 3 are due to the dense sampling by BATSE on CGRO. They provide clear evidence for a time lag between the pulse period evolution and the (O−C) evolution, as already noted by Staubert et al. (2000). A formal cross-correlation of the pulse period residuals and the (O−C) data finds that the (O−C) curve must be shifted to earlier times by ∼35 days to reach maximum correlation, meaning that any developments with time are seen first in the pulse period and then in the timing of turn-on of the next 35 day cycle.

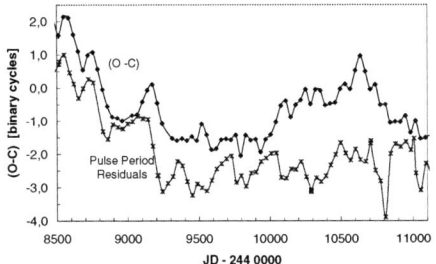 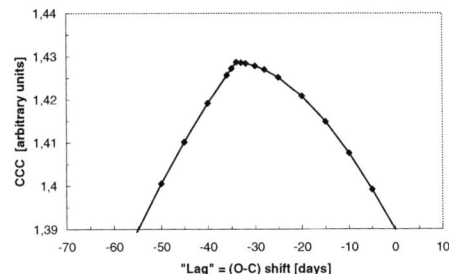

FIGURE 3. Left: (O−C) data (upper curve) and pulse period residuals (shifted down by 1 unit, lower curve) for days 8500-11100 (JD-2440000). Right: Cross correlation coefficient (CCC) between the two curves shown left versus an introduced shift of (O−C). The (O−C) lags behind the pulse period residuals by about 35 days.

(We note that the formal lag value of −35 days must not be over-interpreted, since the resolution of the (O−C) data is, by definition, 35 days. So the likely physical lag is of the order of a few tens of days).

The changes in pulse period are due to transfer of angular momentum to the neutron star. Since the latter should be positively correlated to the mass accretion rate, one might expect to also see a correlation between the pulse period residuals and X-ray flux. While there is some evidence for this (Wilson et al. 1997, Staubert et al. 2000, Still & Boyd 2004), the case is not overwhelming and needs further attention (it will be discussed in Staubert et al. 2006). Further observational parameters are the spectral hardness (Still & Boyd 2004) and the pulse shape (Postnov et al. 2006).

DISCUSSION

A detailed discussion and comparison with models of inclined and warped accretion disks is not possible here. This will appear in Staubert et al. (2006). Here we note that the observed bahavior is consistent both with the Coronal Wind model by Schandl & Meyer (1994) as well as with the Stream-Disk Interaction model by Shakura et al. (1999). Under both models we assume that the motor for the apparent variability is the optical companion which provides more or less material at the inner Lagrangian point. The average mass accretion rate of the NS is such that the system operates close to the equilibrium period with a slight bias towards spin-up. When the mass accretion rate drops also the spin-up rate drops and may even turn to spin-down, in accordance with standard accretion theory (Ghosh & Lamb 1997). The consequences for the observable turn-on times are as observed: within the framework of the coronal wind model (Schandl & Meyer 1994) the reduced X-ray irradiation of the outer parts of the disk reduces the coronal wind and its torque on the accretion disk leading to a generally less inclined disk. This in turn leads to a faster precession of the disk. Anomalous lows are observed when the disks inclination is very small and the view onto the NS is blocked. Within the stream-disk model reduced mass transfer means weaker dynamical action on the disk by the gas stream leading again to a less inclined disk with faster precession. For both models we expect that the NS feels any change in mass transfer first and the response of the accretion disk is delayed by its viscous time scale (a few tens of days) – as observed.

While the short-term behaviour of the discussed observables in Her X-1 are subject to considerable noise in the system, we now believe that the global long-term developments are not due to a random walk (as proposed by Staubert et al. 1983) but rather due to two physical reasons: (1) the optical companion provides *quasi-periodic variations in mass transfer* and (2) the NS itself - through its possible *free precession* (Ketsaris et al. 2000) - provides a stable internal clock, forcing the precession of the disk to stay close to the NS frequency over long time scales, despite the variable forces acting on the disk. The changing mass transfer rate may be due to changing illumination of the optical companion by the X-ray beam. In this sense the apparent 5 year period could be associated with a limit cycle due to a positive feedback in the binary system.

ACKNOWLEDGMENTS

We acknowledge the support by DFG under grants Sta 173/31 and 436 RUS 113/717/0-1 and by the corresponding RBFR grant RFFI-NNIO-03-02-04003 as well as by DLR under grant No. 50 OR 9205. We thank Ljuba Rodina for her valuable contributions to the analysis of the RXTE data, particularly the accurate determination of the pulse periods.

REFERENCES

Bochkarev N.G., et al., 1981, Sov.Astron.Let. 14(6), 421
Boyd P., Still, M., Corbet, R., 2004, ATEL 307
Coburn W., et al., 2000, ApJ 543, 351
Deeter J.E., Boynton P.E., Pravdo S.H., 1981, ApJ 247, 1003
Ghosh P., Lamb F.K., 1979, ApJ 234, 296
Ketsaris N.A., et al., 2000, Proc. "Hot points in Astrophysics", Dubna, p.192
Klochkov D., et al. 2006, A&A, submitted
Nagase F., 1989, PASJ 41, 1
Ögelmann H., et al., 1985, Sp.Sc.Rev. 40, 347
Parmar A.N., Pietsch W., McKechnie S., et al., 1985, Nat 313, 119
Parmar A.N., et al., 1999, A&A 350, L5
Postnov K., et al., 2006, in preparation
Schandl S., Meyer F., 1994, A&A 289, 149
Schandl S., Staubert R., König 1997, Proc. 4th COMPTON Symp., AIP CP 410, 763
Shakura N., et al., 1999, A&A 348, 917
Staubert R., Bezler M., Kendziorra E., 1983, A&A 117, 215
Staubert R., Schandl S., Wilms J., 2000, Proc. 5th COMPTON Symp., AIP CP 510, 153
Staubert R., et al., 2006, to be submitted
Still M., Boyd P., 2004, ApJ 606, L135
Tananbaum H., Gursky H., Kellog E.M., et al., 1972, ApJ 174, L143
Vrtilek S.D., Mihara T., Primini F.A., et al., 1994, ApJ 436, L9
Wilson R.B., Scott D.M., Finger M.H., 1997, Proc. 4th COMPTON Symp., CP 410, 739

Swift X-Ray Telescope Observations of Galactic Transients

Jamie A. Kennea

Department of Astronomy and Astrophysics, 525 Davey Lab, Pennsylvania State University, University Park, PA 16801, USA

Abstract. The *Swift* Gamma Ray Burst Explorer is a multi wavelength satellite mission dedicated to detection and follow-up of Gamma-Ray Bursts. *Swift* is comprised of three instruments, a Gamma-Ray detector (BAT), X-ray telescope (XRT) and UV/Optical telescope (UVOT). *Swift* is able to react with a fast response to both on-board detected transient events, and events discovered by other missions. This rapid response capability makes it an excellent platform for performing time-critical X-ray transients. The *Swift* X-Ray Telescope (XRT) is able to perform imaging observations of X-ray transients and obtain localizations to ~3.5 arc-second accuracy, aiding optical identification with ground based optical and IR follow-ups. We present an account of the operations of the Swift telescope in response to X-ray transients, particularly focusing on XRT observations, and present some results of X-ray monitoring of X-ray Transients with the XRT.

Keywords: X-ray Binaries, Transients, Gamma-Ray Bursts
PACS: 97.30.Qt

INTRODUCTION

The *Swift* Gamma-Ray Burst Explorer[1] was launched on November 20th, 2004, into a low-Earth orbit. The primary mission of the *Swift* mission is to study Gamma-Ray Bursts (GRBs). It performs this task utilizing a combination of multi-wavelength instrumentation and a state of the art attitude control system (ACS) that allows for rapid and accurate slews to targets.

The instrumentation consists of the Burst Alert Telescope (BAT, e.g. [2]), a large area (1.4 sr) coded mask Gamma-Ray telescope primarily utilized to detect GRBs and other transient phenomena; the UV/Optical telescope (UVOT, e.g. [3]), a 30cm modified Ritchey-Chrétien telescope with an image intensified CCD camera covering 170-600nm range and a 17' square field of view, which is primarily utilized for detection of GRB afterglows; and the X-Ray Telescope (XRT), a CCD based imaging X-ray telescope which provides high accuracy (3.5 arc-second error) localization and high resolution X-ray spectra of GRB afterglows [4].

In this paper I will primarily focus of the method of utilization of the *Swift* XRT to perform detection, localization and follow-up observations of Galactic X-ray transients. I also report on results from XRT transient sources.

SWIFT OPERATIONS

In its primary operating mode *Swift* performs rapid follow-up observations of GRBs afterglows, this is performed using a combination of all three instruments. The BAT instrument provides the initial localization of the GRB, usually giving a position accurate to 2.5 arc-minutes. If the target is visible, *Swift* will perform a rapid (~60-300 seconds) slew to the target, and begin observing the source with the XRT and UVOT. The XRT will attempt to find the GRB location automatically by applying a centroiding algorithm to a short (~2.5s) observation of the field. If a position is found this is sent out to the GRB community automatically. The UVOT performs an imaging exposure of the field and Swift science team members send this data to the ground immediately for analysis. After this initial observation *Swift* will continue to observe the GRB field to track the decay of the afterglow. Often this monitoring of GRB afterglows can last from days to months depending on the brightness of the afterglow.

Although GRB science is the primary *Swift* science, its design also makes it an excellent general-purpose telescope. As *Swift* is in a low Earth orbit, only approximately 1/3 of its 96-minute orbit can be utilized to observe a single target. As *Swift* detects GRBs at a rate of approximately 1-2 per week, there is often a large fraction of observing time per orbit that is not utilized for afterglow follow-up. In order to fill this time we perform both "Target of Opportunity" (TOO) and "Fill-in" observations. Fill-in observations are primarily non-time critical observations of interesting science targets proposed by members of the *Swift* science team. These are observed at a low priority and, as their name suggests, are primarily used to fill-in time gaps in the *Swift* observation timeline.

TOO observations are generally considered time critical observations of interesting astrophysical phenomena, and any member of the public via the Swift web page may propose them. TOO observations may be observed utilizing two methods: They can be put into the Swift science plan, which is usually prepared 24 hours before execution or a TOO may be manually uploaded to *Swift* via a telemetry uplink. The latter mode allows for extremely rapid follow-up of time critical events. An uploaded TOO will generally override any pre-planned observations, and TOO uploads typically occur within 1-2 hours of the TOO request, if received during *Swift* office hours. In most cases TOOs of any type are performed within 24 hours of being requested.

The rapid slewing capabilities of *Swift* lead to very high observing efficiencies. In a single orbit *Swift* may observe 3-5 targets. This leads to the possibility of performing very short observations of targets, which would not usually be possible with other X-ray telescopes. TOO requests for observations of bright X-ray/Optical sources for 1 kilo-second are common.

SWIFT XRT AND TRANSIENTS

Although the XRT is primarily designed for localization of GRB X-ray afterglows, it is a fully functional X-ray telescope. The XRT has a field of view of 24 arc-minutes and is able to perform localization with an accuracy of 3.5 arc-seconds. In addition to this XRT has excellent allowing for detection of faint sources to a limiting sensitivity of ~10^{-14} erg/s/cm^2. In addition the XRT produces spectra in the 0.2-10 keV range and is able to detect both continuum and line emission from X-ray sources.

The XRT is generally operated in two read-out modes, Photon Counting (PC) and Window Timing (WT) modes. PC mode provides imaging and spectra of a field, but with a 2.5s timing resolution is highly susceptible to pile-up. WT mode increases the time resolution to ??s, avoiding pile-up, however only 1 dimensional imaging data is produced meaning this cannot be utilized to perform localizations. Typically the XRT will choose the mode in which it operates based on the brightness of sources in the field of view (auto state), however it is also possible for the mode to be fixed in some cases if a certain mode is required, regardless of source brightness (manual state).

The BAT does not detect a large number of X-ray transients, so the primary use of XRT in transient observations is follow up of objects discovered by *INTEGRAL* [5] and *RXTE* [6]. These observations are usually performed with the XRT fixed into a PC mode, to allow for localization. Typical X-ray transients will have XRT count rates of 10-200c/s, meaning only a short (~1ks) observation is required to perform an accurate localization of the source. Figure 1 shows a typical observation of a transient. Often these observations will be followed up with further observations with the XRT in auto state, so good quality spectra may be obtained with zero or much reduced pile-up.

Typically the results of this analysis are quickly reported via the "Astronomers Telegram" (ATEL) [7], in order for the community to perform ground-based optical observations in order to identify the X-ray transient source. In some cases transient sources also have detectable Optical/UV emission as well, meaning *Swift* is able to find a sub arc-second position for the transient. However the lack of red response of the UVOT combined with the fact that X-ray transients are usually highly absorbed means that UVOT detection is relatively rare.

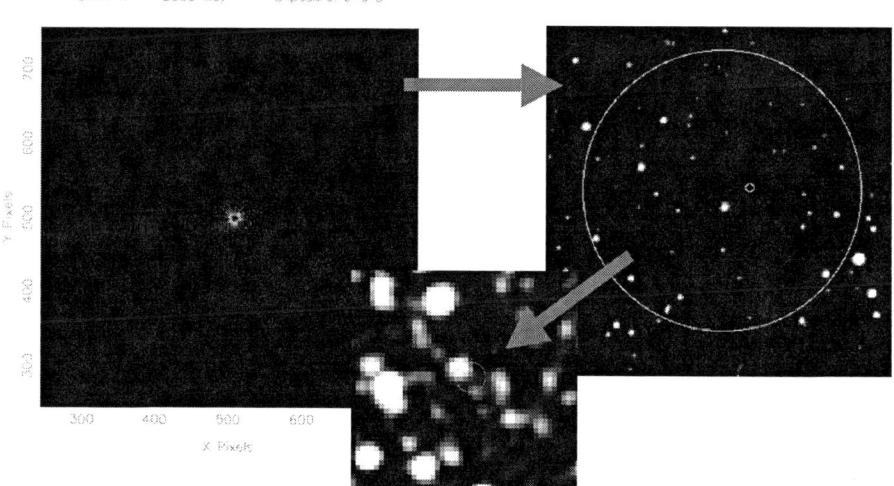

FIGURE 1. Typical observation scenario of a transient source with the *Swift*. The left image shows the XRT image of IGR J17098-3628, note the hole in the PSF is due to pile-up. The image on the right shows the XRT error circle (small) compared to the *INTEGRAL* error circle. The central image shows the XRT error circle on a 2MASS image of the field.

XRT OBSERVATIONS OF TRANSIENTS

XRT has performed many observations of transient X-ray sources during its first year of operations. Here we summarize some highlights from those observations. Table 1 summarizes the XRT localized positions and position errors of observed transient sources, including a reference to the ATEL number in which the position was originally reported.

Swift J1753.45-0127 was a *Swift* discovered transient source. Detected by the BAT [8] on June 30th, 2005, as this target was not a GRB Swift did not perform an automated slew to this target. However on July 1st, 2005, a TOO observation was performed utilizing XRT and UVOT, and an XRT position accurate to 6 arc-seconds was reported [9]. This position was utilized to find an optical counterpart of the source using a ground based observations [10] and the source was later also detected by the UVOT [11].

IGR J16283-4838 was discovered by *INTEGRAL* on April 7th, 2005. *Swift* XRT observations of the source were performed on April 17th and XRT localization allowed the source to be identified in the Spitzer GLIMPSE survey [12]. Follow up observations of this source revealed that the source shows variable absorption (0.7-1.7 x 10^{23} cm^{-2}). Combined observations from *Swift*, *RXTE*, *INTEGRAL* and the *Spitzer* data identified this source as a HMXB, most likely containing a neutron star, embedded in Compton thick material [13].

IGR J16479-4514 was originally detected by *INTEGRAL* in August 2003 and belongs to class of "recurrent fast X-ray transients" recently discovered[14]. This source was detected to flare up by the BAT, and Swift automatically slewed to observe the target with the XRT within 128 seconds of the BAT trigger, on August 30th, 2005. XRT found bright source with a rapidly fading light curve, with some flaring activity (see Figure 2). XRT continued to observe this source for several days where and multiple outbursts from this source were seen during those observations [15].

TABLE 1. XRT localized X-Ray Transients

Name	RA (J2000)	Dec (J2000)	Error (90% conf)	Note	ATEL #
XTE J1701-462	17 00 58.3	-46 11 09.0	6.1	-	781
Swift J174535.5-290135.6	17 45 35.5	-29 01 35.6	6.0	In GC	753
Swift J1626.6-5156	16 26 36.24	-51 56 33.5	3.5	-	688
IGR J01583+5713	01 58 18.2	+67 13 25.9	3.5	Transient	673
IGR J17269-4737	17 26 49.8	-47 38 23.2	8.0	-	626
XTE J1739-286	17 39 54.2	-28 29 44.0	6.0	-	602
Swift J1753.5-0127	17 53 28.3	01 27 09.3	6.0	-	547
HETE J1900.1+2455	19 00 08.4	-24 55 16.0	5.0	ms Pulsar	541
XTE J1747-274	17 47 17.8	-27 20 30.0	6.0		500
IGR J17098-3626	17 09 45.9	-36 27 57.0	5.0	-	476
IGR J16283-4838	16 28 10.7	-48 38 55.0	5.0	See ref [13]	456

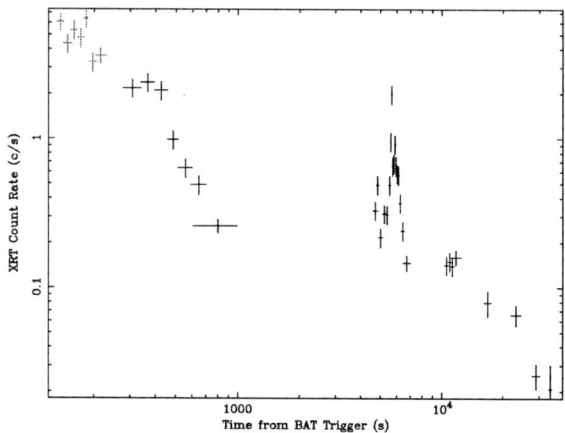

FIGURE 2. Swift XRT lightcurve of IGR J16479-4515.

CONCLUSION

With its excellent positional accuracy, spectral response and sensitivity the XRT is an ideal instrument with which to perform follow-up observations of X-ray transients. This is further enhanced by the ability of *Swift* to perform rapid follow-up observations of transient sources often within hours of their discovery, with high observing efficiency allowing for short observations to be taken where only a position is needed. TOO requests are open to any member of the community, and outside participation is actively encouraged.

Swift and the XRT provide an ideal tool for the rapid localization and long term follow-up of X-ray transients. This has been proven by the very active role that *Swift* has taken in its first year of observations in providing follow-up observations of transients.

REFERENCES

1. N. Gehrels et al., *Astrophysical Journal*, 2004, **661**, pp. 1005.
2. S. D. Barthelmy et al., *Space Science Reviews*, 2005, **120**, pp. 143.
3. P. W. A. Roming et al., *Space Science Reviews*, 2005, **120**, pp. 95.
4. D. N. Burrows et al., *Space Science Reviews*, 2005, **120**, pp. 165.
5. C. Winkler et al., *Astronomy and Astrophysics*, 2003, **411**, pp. L1-L6.
6. H. V. Bradt et al., *Astronomy and Astrophysics Suppl. Series*, 1993, **97**, pp. 355-360.
7. R. E. Rutledge, *Publications of the Astronomical Soc. of the Pacific*, 1998, **110**, pp. 754-756.
8. D. M. Palmer et al., *Astronomers Telegram*, 2005, #546.
9. D. N. Burrows et al., *Astronomers Telegram*, 2005, #547.
10. J. P. Halpern, *Astronomers Telegram*, 2005, #549.
11. M. Still, *Astronomers Telegram*, 2005, #553.
12. R. A. Benjamin et al., *Publications of thee Astronomical Soc. of the Pacific*, **115**, pp. 953.
13. V. Beckmann et al., *Astrophyical Journal*, **631**, pp. 506.
14. A. Sguera et al., *Astronomy and Astrophysics*, **446**, pp 471.
15. J. A. Kennea et al., *Astronomers Telegram*, 2005, #599.

Can Black Holes Provide the Emitted Energy of GRBs?

Reuven Opher

*IAG, Universidade de São Paulo, Rua do Matão 1226,
Cidade Universitária, CEP 05508-900 São Paulo, SP, Brazil*

Abstract. Charged or rotating black holes (BHs) resulting from the collapse of the cores of massive stars can, in principle, be a source of considerable energy. This energy source has been suggested as the origin of gamma-ray bursts. I show that the plasma density around such BHs is always sufficiently high so as to make the suggested mechanisms for energy extraction inviable.

Keywords: physics of black holes — gamma-ray bursts — gamma-ray — relativity
PACS: 04.70.-s, 98.70.Rz, 95.85.Pw, 95.30.Sf

1. INTRODUCTION

Charged or rotating black holes (BHs) are often cited as the origin of gamma-ray bursts (GRBs). In order for the production of GRBs due to the direct extraction of the energy of BHs (i.e., without including an accretion disk in the model) to actually take place, it must occur in realistic astrophysical conditions. BHs are formed in the collapse of the cores of massive stars ($\gg M_\odot$) of radii $\sim 10^9$ cm. The density of these cores is extremely high, much greater than that of the Sun, ~ 100 $g\ cm^{-3}$. I show here that in such high densities, a charged black hole (CBH), which was studied by Damour and Ruffini [1], cannot be used as a source for GRBs. At such densities, the energy extracted from a rotating (RBH) by the Blandford-Znajek mechanism (BZM) [2] is also shown to be insufficient to produce GRBs.

In Section 2, I discuss the extraction of energy from a CBH and in Section 3, the extraction of energy from a RBH. Discharging the CBH is discussed in Section 4 and short circuiting the potential of a RBH, in Section 5. Conclusions are presented in Section 6.

2. EXTRACTING ENERGY FROM A CBH

According to the Darmour-Ruffini theory [1], $1.8 \times 10^{54}(M_{BH}/M_\odot)$ ergs can be extracted from CBHs with masses $3.2 M_\odot \leq M_{BH} \leq 7.2 M_\odot$, which is sufficient to explain GRBs. The total energy of a BH, E_{BH}, with a charge Q and an angular momentum J is

$$E_{BH}^2 = \left(M_{ir} c^2 + \frac{Q^2}{2\rho_+} \right)^2 = \frac{J^2 c^2}{\rho_+^2}, \tag{1}$$

where
$$\rho_+^2 \equiv 4\left(\frac{G^2}{c^4}\right)M_{ir}^2 \tag{2}$$
and M_{ir} is the irreducible mass of the BH. The combined upper limit on Q and J is
$$Q^4 + 4J^2c^2 \leq \left(\frac{\rho_+^2 c^4}{G}\right)^2 \leq (4G)^2 M_{ir}^4. \tag{3}$$
For a maximally charged BH (MCBH),
$$\frac{Q}{M_{ir}\sqrt{G}} \simeq 1. \tag{4}$$
Since, for a proton,
$$\frac{q_p}{m_p\sqrt{G}} \simeq 1 \times 10^{18}, \tag{5}$$
a MCBH could have conceivably been constructed from a hydrogen mass, in which for every 10^{18} hydrogen atoms, one additional proton is present. Problems with energy extraction from a MCBH to explain GRBs are discussed in Section 4.

3. EXTRACTING ENERGY FROM A RBH

A detailed analysis of the extraction of energy from a RBH to explain GRBs was made by Lee, Wijers, and Brown [3]. The mechanism of the energy extraction is based on the theory of Blandford and Znajek [2], who showed that all RBHs in a magnetic field act as rotating conducting shells with an effective internal resistance $R_{INT} = 377$ ohms. Lee et al. suggested that the energy of the current circuit is deposited into a distant load, far from the RBH. They showed that the energy available for a GRB is $1.6 \times 10^{53}(M_{BH}/M_\odot)$ ergs. In order to extract the energy from the RBH in times ≤ 1000 s, necessary to explain GRBs, a magnetic field $\sim 10^{15}$ G near the RBH is required.

The energy of rotation of a RBH is
$$E_{RBH} = f(a)M_{BH}c^2, \tag{6}$$
where
$$a \equiv Jc/M_{BH}^2 G \tag{7}$$
and
$$f(a) = 1 - \left[\frac{1}{2}\left(1+\sqrt{1-a^2}\right)\right]^{1/2} < 0.29. \tag{8}$$
The luminosity L_{RBH} emitted is
$$L_{RBH} = 6.7 \times 10^{50}\left(\frac{B}{10^{15}G}\right)^2 \left(\frac{M_{BH}}{M_\odot}\right)^2 \text{ erg s}^{-1}. \tag{9}$$

This luminosity appears to be able to explain GRBs. Problems connected with this model are discussed in Section 5.

4. DISCHARGING A CBH

A CBH (e.g., a BH which is positively charged) immersed in an astrophysical plasma will naturally tend to discharge as electrons are attracted to it and ions are repulsed. Thus, the charge of a CBH would be neutralized if the plasma density around it were to be sufficiently high.

The radius of the core of a massive star is $\sim 10^9$ cm before it collapses into a BH. The number density n within a distance $\sim 10^9$ cm from the center of the sun, is

$$n \sim 10^{26} \text{ cm}^{-3}. \tag{10}$$

When a massive stellar core, $3 - 7 M_\odot$, collapses to form the CBH,

$$n \gg 10^{26} \text{ cm}^{-3}. \tag{11}$$

Let us evaluate the density of the plasma required in a spherical shell of radius $R \sim 10^9$ cm with a thickness ΔR in order to discharge a CBH. The number of electrons required to discharge a 1/10 maximally charged BH is

$$N \sim 10^{39} \tag{12}$$

The density of the charges in the shell is $N/4\pi R^2 \Delta R$. For $\Delta R \sim 0.1R$,

$$n \sim 10^{12} \text{ cm}^{-3}. \tag{13}$$

This density is typical of the solar corona and not the interior of a star.

We assumed in the above that the electron mean free path is greater than the shell thickness ΔR. Let us see if this is reasonable.

For a nonrelativistic plasma, the collision frequency v_c is

$$v_c = n \left[\frac{\ln \Lambda}{0.267 \, T^{3/2}} \right], \tag{14}$$

where $\ln \Lambda$ is the Coulomb logarithm, which has a weak dependence on density and temperature. For the densities discussed here (see Tables 1 and 2), $\ln \Lambda \sim 20$.

However, the realistic plasma near a black hole is, in fact, relativistic, with a temperature $\sim 2m_ec^2/k_B \sim 10^{10}$ K. For such a plasma, the electrons, positrons, and photons are in quase equilibrium. Electron-photon scattering dominates, with a cross section $\sigma_T \simeq 10^{-24}$ cm^2. The mean free path λ is then

$$\lambda \simeq \frac{1}{n\sigma_T}. \tag{15}$$

Using Eq. (14), it is found that at a temperature $T \sim 10^{10}$ K, $\lambda \simeq c/v_c$ is approximately that of Eq. (15). Therefore, we use Eq. (15) for the mean free path for the electrons at the temperature $T \sim 10^{10}$ K.

TABLE 1. The parameters n, tg, λ, and τ vs ΔR for a CBH

ΔR cm	n cm^{-3}	tg s	λ cm	τ s
10^8	$\sim 10^{12}$	$\sim 10^8$	$\sim 10^{12}$	$\sim 10^{-3}$
10^5	$\sim 10^{15}$	$\sim 10^7$	$\sim 10^9$	$\sim 10^{-6}$
1	$\sim 10^{20}$	$\sim 10^4$	$\sim 10^4$	$\sim 10^{-11}$

For a sphere of uniform density, the gravitational collapse time depends only on the density. The gravitational collapse time t_g for a hydrogen-like plasma is

$$t_g = \frac{4 \times 10^{14}}{n^{1/2}} \text{ s}, \tag{16}$$

where n is the particle number density. Various values for ΔR, n, t_g, λ, and τ are shown in Table 1. The characteristic time for the relativistic electrons to cross the shell thickness is $\tau = \Delta R/c$.

It can be observed in Table 1 that in all cases, the mean free path of the electrons is four orders of magnitude greater than the shell thickness. We also note that the densities in Table 1 that are needed to neutralize a CBH, are much less than the expected density in Eq. 11. The crossing times τ in Table 1 are also much shorter than the gravitational collapse times.

5. SHORT-CIRCUITING THE POTENTIAL OF A RBH

A RBH, which develops a potential difference between its poles of rotation due to the presence of a magnetic field, tends to be neutralized by the ambient plasma. Electrons are attracted to the positive pole and ions, to the negative pole. Let us evaluate the number of charges needed to neutralize and effectively short-circuit the potential difference of a RBH.

To explain GRBs, we need to produce $\sim 10^{52}$ ergs, which requires a current $I \sim 10^{21}$ A in a time $t \sim 5$ s. The number of charges transfered is

$$N \sim \frac{I \times t}{e} \simeq 3 \times 10^{40}. \tag{17}$$

We note that this is a factor of ten larger that the number of charges required to neutralize a CBH (Eq. (12)). Thus the discussion of Section 4 can be repeated here. The only difference is that instead of the values in Table 1, we now have the values shown in Table 2.

Since these densities are much lower than that in Eq. (11), the ambient plasma would short-circuit any potential difference created between the poles of a RBH very rapidly, as indicated by the short τ times in Table 2. Consequently, the energy that was to be extracted by the BZM in producing an electric current and depositing its energy in a

TABLE 2. The parameters n, tg, λ, and τ vs ΔR for a RBH

ΔR cm	n cm^{-3}	tg s	λ cm	τ s
10^8	$\sim 10^{13}$	$\sim 10^8$	$\sim 10^{11}$	$\sim 10^{-3}$
10^5	$\sim 10^{16}$	$\sim 10^6$	$\sim 10^8$	$\sim 10^{-6}$
1	$\sim 10^{21}$	$\sim 10^4$	$\sim 10^3$	$\sim 10^{-11}$

load far from the RBH, would actually be deposited right next to the RBH. Therefore, no effective energy extraction would have taken place and no GRBs would have been produced.

6. CONCLUSIONS

The direct extraction of energy from BHs to power GRBs is certainly a very attractive idea. However, I showed here that ambient plasma densities near the BHs, formed in the collapse of massive stars, are always sufficiently high so that the mechanisms suggested for direct energy extraction are, in fact, inviable. The plasma densities required in order to discharge a CBH or short circuit the potential of a RBH are very much lower than those in stellar interiors. If, in fact, BHs are involved in the creation of GRBs, they do not do so in the manner described above, but likely through a mechanism which is connected with the accretion disk.

ACKNOWLEDGMENTS

The author would like to thank the Brazilian agencies CNPq (300414/82-0) and FAPESP (00/06770-2) for partial support.

REFERENCES

1. T. Damour and R. Ruffini, *Phys. Rev. Lett.*, **35**, 463 (1975).
2. R. D. Blandford and R. L. Znajek, *Mon. Not. R. Astron. Soc.*, **179**, 433 (1977).
3. H. K. Lee, R. A. M. J. Wijers, and G. E. Brown, *Phys. Rep.*, **325**, 83 (2000).

MIRAX SCIENCE—MULTIWAVELENGTH

Jets from Galactic X-ray Transients the MIRAX Perspective

Elena Gallo* and Rob Fender[†]

*University of California Santa Barbara, Physics Department, CA 93106-9530, U.S.
[†]School of Physics & Astronomy, University of Southampton, Highfield, SO17 1BJ, U.K.

Abstract. In recent years, coordinated radio and X-ray observations of Galactic relativistic jet sources have led us to a fairly good phenomenological understanding of the jet and accretion properties over different accretion regimes. While there are well established classifications of these sources in terms of X-ray spectral and timing features, and appearance of the radio core, the difficulty comes about in assessing the interactions between the inflow and the outflow, and in understanding the physics of the transition between the different regimes. In light of these issues, we briefly review the radio properties of black hole and neutron star X-ray binaries with a focus on specific areas in which the Brazilian X-ray astronomy satellite MIRAX, combined with pointed radio observations, will enable us to gain insights into the time variable jet-accretion coupling and related phenomena.

Keywords: <Enter Keywords here>
PACS: 97.80.Jp; 95.58.Fd

BLACK HOLE X-RAY BINARY JETS

Time-Variable Radio Properties

The largest body of observational data that pertain to Galactic relativistic jet sources is associated with black hole X-ray binaries (BHBs). We refer the reader to Remillard, these Proceedings, for a description of the spectral and timing X-ray properties of BHBs over the (five) different X-ray states (after [1]), and to [2] for a complete review on X-ray binary jets. Multi-wavelength observations of BHBs have established a relatively clear pattern of behaviour; specifically:

- hard state sources are associated with persistent radio emission and flat/inverted spectra which extend up to the mid-IR band [3]. The flat spectra are modelled as a superimposition of synchrotron peaks coming from different regions along a partially self-absorbed conical outflow [4], often referred to as 'steady jet'. While the hard state core radio emission is generally unresolved, VLBA observations have confirmed the jet-interpretation in two sources, imaging collimated radio sources on milliarcsec-scales (Cyg X-1 [5]; GRS 1915+105 [6]).
- thermal dominant X-ray states are associated with no detectable core radio emission; the radio fluxes are quenched by a factor ~ 50 with respect to the hard state (e.g. [7]). probably implying the physical suppression of the jet in this regime.
- bright radio flares are often observed during hard-to-soft state transitions, and are thought to be the signature of powerful ejection events.

CP840, *The Transient Milky Way: A Perspective for MIRAX*,
edited by F. D'Amico, J. Braga, and R. E. Rothschild
© 2006 American Institute of Physics 0-7354-0332-5/06/$23.00

- following the state transition, large scale (several arcsec), fading radio plasmons are seen moving away from the binary core with highly relativistic velocities.

There is mounting evidence that the radiatively inefficient hard (and quiescent) state may be a 'jet-dominated' regime, in the sense that more energy is carried off in a jet (in the form of internal energy, bulk flow, and radiation) than is emitted by the disc-corona and goes through the horizon (disc-corona internal energy, bulk flow, and emission). We refer the reader to [8], [9], [10] for more details, and to [11] for a scenario in which the radiative output from the jet alone would excess the disc-corona energy budget.

A Unified Model for State Transition and Jet Production

[12] have attempted to construct a unified, phenomenological model for the disc-jet coupling in BHBs. The model can be briefly summarised as follows: the source is in the low-luminosity hard X-ray state, producing a steady jet. At some point the X-ray luminosity starts to increase while the X-ray spectrum softens; the steady jet persists during this unstable phase. The X-ray spectrum keeps softening, to be interpreted as the accretion disc inner boundary moving closer and closer to the BH. As a consequence, the jet properties change, most notably its velocity (as the escape velocity from the inner region is increasing). A higher Lorentz factor 'plasmon' is ejected, causing the propagation of an internal shock through the slower-moving outflow in front of it, giving rise to a bright radio flare. Eventually, the result of this shock is what we observe as a post-outburst, optically thin discrete ejection (see [13]). The bright radio flare associated with the X-ray state transition could coincide with the very moment in which the hot corona of thermal electrons, likely to be responsible for the X-ray power law in the spectra of hard state BHBs, is accelerated and ultimately evacuated. Figure 1 illustrates an example of a powerful radio flare in a prototypical BHB.

The continuous hard X-ray coverage (10-100 keV) provided by the CXD (Cameras de Raios-X Duros) on-board MIRAX, coordinated with pointed radio observations, will enable us to establish a causal connection, if any, between the softening of the hard X-ray spectrum and the radio properties. To illustrate the vast potentiality of a continuous hard X-ray monitoring campaign, it is worth mentioning that recent triggered RXTE/INTEGRAL observations of the BHB GX339–4 in outburst [14] have shown that the high energy ($\sim$100 keV) cutoff typical of hard state X-ray spectra, either disappears or shifts towards much higher energies within timescales of hours (<8 hr) during the transition. Previous suggestions of such behaviour were based on X-ray monitoring campaigns with instruments such as OSSE, for which the long integration times required in order to accumulate significant statistics did not allow to constrain the timing and significance of rapid changes in the X-ray spetcra. Radio observations of the same source, conducted during the rise of a previous outburst [15], and of GRS1915+105 ([12]), indicate that in this phase, and prior to the bright radio flare, the jet spectral index seems to 'oscillate' in an odd fashion, from flat to inverted to optically thin, as if the jet was experiencing some kind of instability as the X-ray spectrum softens. The post-radio flare phase, on the other hand, seems to be related with a sharp change in the X-ray

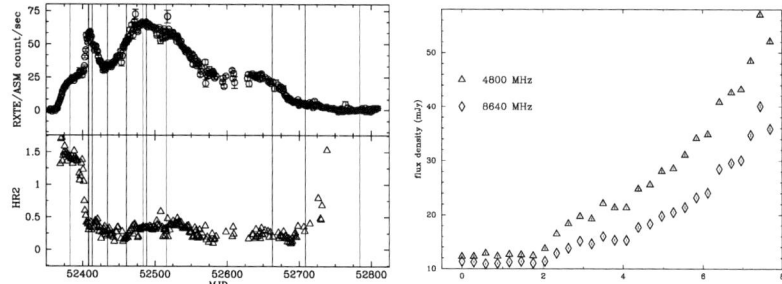

FIGURE 1. *Left panel*: X-ray light curve of the BHB GX 339–4 over its 2002 outburst (top), and hardness ratio (bottom), as observed by the RXTE ASM. The vertical lines indicate the dates of the pointed radio observations. *Right panel*: dual frequency radio light curve of GX 339–4 almost simultaneously with the first peak in the X-ray light curve, indicated by the thick vertical line in the left panel. Radio observations at later times imaged a fading relativistic radio jet on arcsec scales, most likely powered by the flare. Simultaneus hard X-ray monitoring of a similar event, by MIRAX CXD, will provide us with useul information on the timing of the sudden rise in radio flux and the sudden changes in the X-ray spectral properties.

timing properties of the systems: [16] identifies the flaring with the hard limit of the so called steep power law state.

Steering almost permanently at the Galactic Plane with a factor 10 better sensitivity than the All Sky Monitor on-board the RXTE and with continuous hard X-ray coverage provided by CXD, MIRAX will be enable us to perform systematic studies of the yet poorly understood X-ray state transitions. Ideally one would like to have hard X-ray light curves simultaneous with those of the bright radio flares that accompany the X-ray state transitions, in order to establish what drives what, and what is the relation of the high energy power law with the radio emission.

NEUTRON STAR X-RAY BINARY JETS

A comparison with black holes

In terms of radio properties, the picture for neutron star X-ray binaries is not as complete as for the BHs; a recent comprehensive study comparing the two classes of sources is presented in [17]. There are clear qualitative similarities: as in BHBs, hard states below about 1 per cent of the Eddington luminosity are associated with persistent unresolved radio emission, while outbursting and variable sources at the highest luminosities are known to power optically thin radio ejecta displaying relativistic motions away from the binary core on arcsec-scales. However, there are important quantitative differences, the most important being that NSs are less radio loud (see also [18]) for a given X-ray luminosity (by a factor of about 30) and they do not show the strong suppression of radio emission in the thermal dominant state that is observed in BH systems. This suggests that a property that is unique to the nature of the accretor

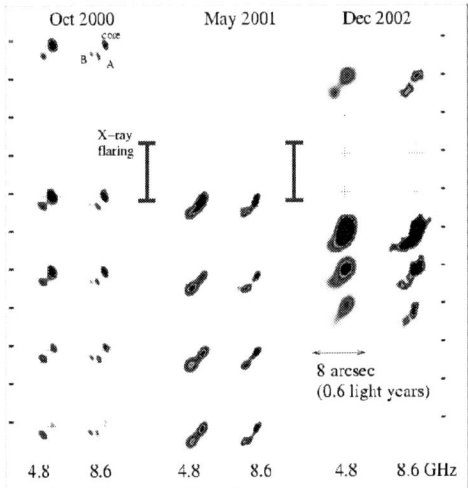

FIGURE 2. Sequences of dual frequency radio observations of the NS X-ray binary and Z source Cir X-1. The binary core position is marked by a cross. The observations reveal that following the X-ray flaring in this source, the extended radio structure brightens on timescales of days. The apparent velocities exceed 10c, indicating a NS accretor is capable of powering ultrarelativistic jets. From [20]. This source is also among the Z sources for which transient hard X-ray tails have been detected [21], and for which a jet-origin has been suggested by some authors.

might somehow regulate the jet power, most notably the BH spin could boost it the or the NS magnetosphere could halt it (or a combination of the two).

At the same time however, radio observations of the NS X-ray binary Cir X-1[1] indicates that this system is capable of powering ultra-relativistic outflows [20], with bulk Lorentz factors in excess of 10 (see Figure 2). These observations prove that the production of highly relativistic jets does not require properties that are unique to BHs, such as an event horizon or extreme spin values, thereby effectively disproving the so called 'escape-velocity' paradigm.

Hard tails

The class of Z sources comprises 6 NS X-ray binaries (possibly 7, with the recent addition of XTE J1701–462; [22]) emitting at or near the Eddington limit. The designation Z source results from the shape they trace as the move in the X-ray colour-colour diagram, probably as the result of changes in the prompt vs. average mass accretion rate [23]. Transient hard X-ray tails extending up to 100 keV have been discovered from classical Z sources and from the peculiar Z source Cir X-1 ([24] and references therein).

---

[1] The (previously debated) nature of the compact object is this system has been recently established by the discovery of twin kHz quasi-periodic oscillations [19].

The time behaviour of the hard tail is complex and still not well understood. The X-ray spectra of Z sources resemble those of BHBs in the thermal dominant state, suggesting a common origin of the hard X-ray power laws in these systems. As all the Z sources are detected as bright and variable radio sources, an interesting possibility is that the non-thermal, high energy electrons responsible for the hard tails observed in Z sources might originate in a jet [25], either because the relativistic electrons that are accelerated off the binary core in the jet are part of the population that inverse-Compton scatters the thermal disc photons, or, in a more radical interpretation, because the hard X-ray tail is actually the optically thin power law tail of the jet synchrotron spectrum.

Once again, the continuous hard X-ray coverage of MIRAX CXD will enable us to perform systematic studies of those hard tails, which, together with pointed radio observations could help to discriminate between the various emission processes (note that, with the exception of Sco X-1 and Cyg X-2, all the Z sources are located close to the Galactic Plane).

ACKNOWLEDGMENTS

E.G. gratefully acknowledges financial support from the National Institute for Space Research in Brazil, and wishes to thank for the warm hospitality in São José dos Campos.

REFERENCES

1. J. E. McClintock and R. A. Remillard, "Black Hole X-ray Binaries" in *Compact Stellar X-ray Sources*, edited by W. Lewin and M. van der Klis, Cambridge University Press, Cambridge, 2005
2. R. Fender, "Jets from X-ray Binaries" in *Compact Stellar X-ray Sources*, edited by W. Lewin and M. van der Klis, Cambridge University Press, Cambridge, 2005
3. R. Fender, *Mon. Not. Roy. Astron. Soc.*, **322**, 31-42 (2001).
4. R. D. Blandford and A. Königl, *Astr. J.*, **232**, 34-48 (1979)
5. A. M. Stirling et al. , *Mon. Not. Roy. Astron. Soc.*, **327**, 1273-1278 (2001)
6. V. Dhawan, I. F. Mirabel and L. F. Rodríguez, *Astr. J.*,**543**, 373-385 (2000)
7. R. Fender et al. , *Astr. J.*,**519**, L165-L168 (1999)
8. R. Fender, E. Gallo and P. Jonker, *Mon. Not. Roy. Astron. Soc.*, **343**, L99-L103 (2003)
9. M. Livio, J. E. Pringle and A. R. King, *Astr. J.*, **593**, 184-188 (2003)
10. J. Malzac, A. Merloni and A. C. Fabian, *Mon. Not. Roy. Astron. Soc.*, **351**, 253-264 (2004)
11. S. Markoff, H. Falcke and R. Fender, *Astron. Astrophys*, **372**, L25-L28 (2001)
12. R. Fender, T. M, Belloni and E. Gallo, *Mon. Not. Roy. Astron. Soc.*, **355**, 1105-1118 (2004)
13. C. Kaiser and R. Sunyaev, *Astron. Astrophys*, **356**, 975-988 (2000)
14. T. M. Belloni et al. , *Mon. Not. Roy. Astron. Soc.*, in the press (2006)
15. E. Gallo et al. , *Mon. Not. Roy. Astron. Soc.*, **347**, L52-L56 (2004)
16. R. Remillard, "X-ray States of Black-Hole Binaries and Implications for the Mechanism of Steady Jets", in Texas@Stanford 2004, edited by P. Chen, SLAC Electronic Conference Proceedings Archive
17. S. Migliari and R. Fender, *Mon. Not. Roy. Astron. Soc.*, **366**, 79-91 (2006)
18. R. Fender and M. Hendry, *Mon. Not. Roy. Astron. Soc.*, **317**, 1-8 (2000)
19. S. Boutloukos, R. Wijnands and M. van der Klis, *The Astronomer's Telegram*, 695 (2006)
20. R. Fender et al. , *Nat*, **427**, 222-224 (2004)
21. R. Iaria et al. , *Astr. J.*, **547**, 412-419 (2001)
22. J. Homan et al. , *The Astronomer's Telegram*, 725, (2006)
23. M. van der Klis, *Astr. J.*, **561**, 943-949 (2001)
24. T. Di Salvo, L. Stella and N. Robba, *Mem. Soc. Astron. Ital.*, **73**, 1082-1087 (2002)
25. T. Di Salvo et al. , *Astr. J.*, **544**, L119-L122 (2000)

Multiwavelength Variability in Transient Black Hole Binaries

Robert I. Hynes

Department of Physics and Astronomy, Louisiana State University, Baton Rouge, Louisiana 70803, USA; rih@phys.lsu.ed

Abstract. Variability is one of the ubiquitous characteristics of accreting black holes extending across all wavelengths. Variability provides information about timescales and causal connections within and between components of the accretion flow and/or outflow, and several independent mechanisms appear to be working to produce correlations between wavelengths. One mechanism is reprocessing of X-ray variability by the disk and/or companion star, another appears to be related to emission from a jet, or the accretion flow at the base of a jet. This brief review will summarize results from two outbursts of black hole candidate transients this year, XTE J1118+480 and Swift J1753.5–0127.

Keywords: accretion, accretion disks—binaries: close—stars: individual: XTE J1118+480, Swift J1753.5–0127
PACS: 97.80.Jp

INTRODUCTION

X-ray binaries, and more specifically the transient black hole binaries that are the focus of this contribution, are variable objects. This is a universal characteristic seen from the radio through optical to X-rays and γ-rays. The X-ray variability is usually dominated by instabilities in the accretion flow. X-rays irradiate the outer accretion disk and companion star, resulting in reprocessed optical and UV radiation which is expected to be imprinted with the same variability as the X-ray signal is. An important difference, however, is that the optical and X-rays originate from a volume of significant spatial extent, resulting in light travel time delays between the X-rays and the reprocessed emission. It is then possible to infer information about the geometry and scale of the reprocessing region from the lags measured between X-ray and optical/UV variability; this technique is known as reverberation or echo-mapping, as the reprocessed light behaves as an echo [16, 5]. In attempting to apply the technique it has also emerged that not all optical/UV correlations arise in reprocessing at all; in some cases, an alternative kind of correlations are present that may originate in optical/UV synchrotron emission from a jet [4].

In this work we will summarise key results of variability studies of two transient sources that went into outburst in 2005. These show both of these kinds of correlated variability, with correlations in XTE J1118+480 appearing to be dominated by optical/IR jet emission and Swift J1753.5–0127 appearing consistent with thermal reprocessing.

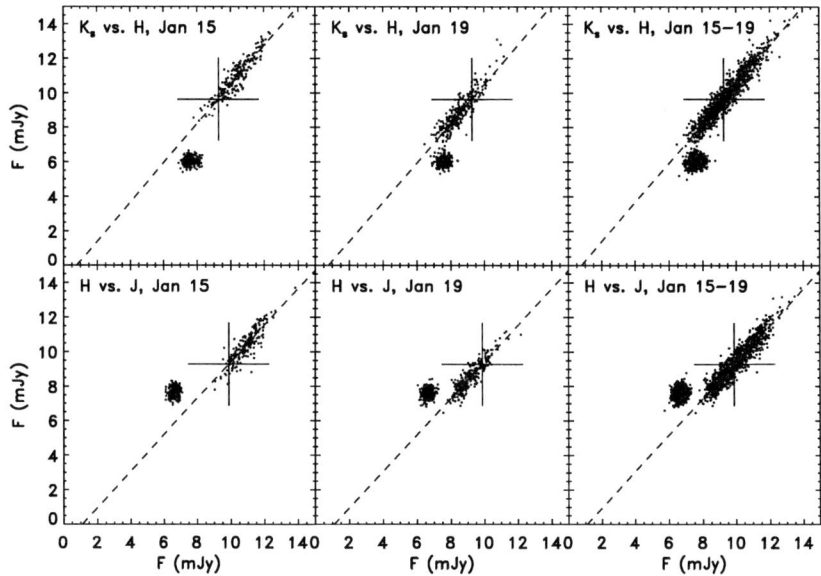

FIGURE 1. J vs. H and H vs. K_s fluxes of XTE J1118+480 for the first night, the last night, and all nights with 2 s data combined. Two clusters of points are present on each plot. The diagonally dispersed ones are XTE J1118+480, the circular clump are the comparison star. The dashed lines all show the same linear fit to the combined dataset (for that color), to provide a point of reference in comparing the first and last night. The cross indicates the run averaged fluxes as a point of reference.

XTE J1118+480

The X-ray source XTE J1118+480, and its optical counterpart KV UMa, were discovered during an outburst in 2000 [18]. Optical spectroscopy in quiescence established a mass function of $\sim 6 M_\odot$, confirming that this system harbors a black hole [11, 20]. It is in a fortunate location of very low Galactic extinction permitting the most comprehensive multiwavelength coverage ever possible for a black hole binary [3, 12, 1]. Modeling of this spectral energy distribution has raised the possibility that much of the IR and even optical emission is arising in synchrotron emission from a jet [13, 21].

Additional evidence in favor of optical/IR synchrotron emission came from observations of rapid correlated X-ray and optical/UV variability [10, 4]. These showed excessively high brightness temperatures for thermal emission and cross-correlation and lag properties inconsistent with expectations of reprocessing in the disk and/or companion star. Surprisingly the strength of the variability increased from a few percent in the UV, to $\sim 10\%$ in the optical and 30–50% in the near-IR. This is also inconsistent with expectations of reprocessing in hot X-ray heated areas. The spectral energy distribution of the variability implied by this was closer to that of optically thin synchrotron emission, although this was obtained by combining non-simultaneous measurements of the r.m.s.

variability.

XTE J1118+480 went into outburst again in early 2005 [22]. On this occasion we were in the fortunate position of already being on the KPNO 2.1 m telescope using the SQIID camera, a multi-channel imager able to take J, H, and K_s images simultaneously; an ideal instrument to explore the spectrum of the variability in this source [6]. These results are described fully by Hynes et al. [9]; we here summarize the main conclusions with respect to the origin of the IR variability. Observations were obtained on five nights using either 1 s or 2 s exposures in J, H, and K_s simultaneously. The deadtime between observations was large, ~ 54 s, so we do not obtain useful lightcurves. Each observation does, however, provide a snapshot of the spectral energy distribution (SED). By comparing these we can isolate the SED of the variable component from that of persistent emission, resolving an uncertainty in attempts to model the average SED. Not surprisingly, we find good correlations between bands, and as in the previous outburst the IR variability spans a large range of fluxes, dominating over statistical uncertainties (Fig. 1). The slope of the correlation measures the color or spectrum of the variable component, for example dF_K/dF_H translates into the $H - K$ color. Maximum spectral leverage is provided by combining all three bands and in this range we find a power-law fit to the implied spectrum is as good as any other. Multiple nights of coverage provide consistency checks and excellent repeatability from night to night. As in the previous outburst, we find that the variability is very red, and the variability SED we deduce corresponds to a power-law $F_\nu \propto \nu^\alpha$, where $\alpha = -0.78 \pm 0.07$. This is in excellent agreement with expectations of optically thin synchrotron radiation from a jet [13], but cannot be reconciled with thermal emission, given the red spectrum and short variability time-scales. These observations provide further evidence for synchrotron emission from a jet in the IR in this object.

SWIFT J1753.5–0127

Another X-ray transient outburst occurred in 2005 June in the form of Swift J1753.5–0127, the first Swift-discovered X-ray transient [17]. X-ray timing and spectral characteristics were typical of black hole candidates [14, 15]. An optical counterpart was identified by Halpern [2] and confirmed to be variable on short timescales [7] prompting a search for correlated variability.

We were again fortunate, as the optical observations were kindly obtained by F. Mullally, who was observing with the Argos CCD Photometer on the McDonald Observatory 2.1 m telescope. Two observations were able to be coordinated with *RXTE* public observations providing simultaneous coverage on 2005 July 6 and 7. The Argos camera was operated in 1 s time-resolution mode to obtain the maximum resolution possible of lags between X-ray and optical. Unlike in the case of XTE J1118+480, the characteristics of the optical variability, for example its auto-correlation function as compared to the X-ray one, were as might be expected for thermal reprocessing, consistent with the disk-like optical/UV SED observed by Still et al. [19]. We show in Fig. 2 an example of a cross-correlation function obtained on 2005 July 6. The prompt peak followed by an extended tail is as would be expected for reprocessing in an accretion disk [16]. The lack of an extended response points to a relatively compact reprocessing region. If the

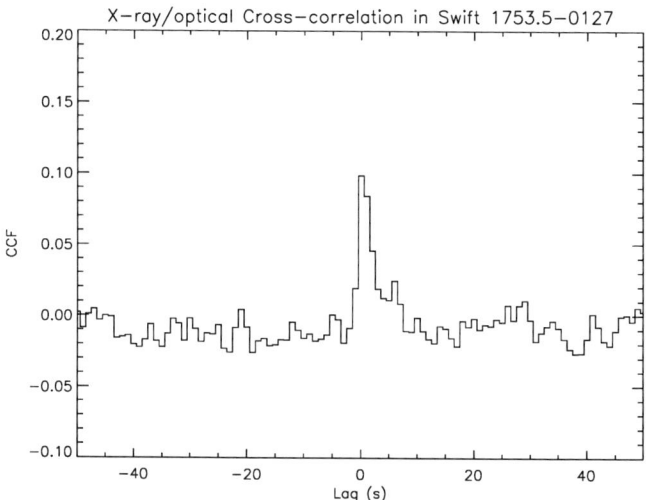

FIGURE 2. Cross-correlation function between X-ray and optical data of Swift J1753.5–0127 from 2005 July 6. A positive lag indicates the optical lagging behind the X-rays following common convention. A very similar cross-correlation was obtained on the following night.

response is distributed over the disk, then the range of lags present implies that this object falls among the typical short-period black hole transients, with an orbital period of half a day or less.

CONCLUSIONS

Correlated X-ray and UV/optical/IR variability during X-ray transient outbursts is a common feature. Its interpretation is not obvious without inspection of the data carefully, however. In some cases, as here for Swift J1753.5–0127, correlated variability appears to follow the standard paradigm for prompt thermal reprocessing of X-rays with the only lags arising from light travel times within the system. In other cases, however, the correlations appear to arise from variability in direct optical/IR emission from a synchrotron jet. XTE J1118+480 is the best example of this case.

ACKNOWLEDGMENTS

The author's participation in this meeting was funded by a Louisiana State University Council on Research Faculty Travel Grant. The observations of Swift J1753.5–0127 described were made possible thanks to Fergal Mullally for sacrificing some of his McDonald time to observe this object, and to Jean Swank and the *RXTE* team for accomodating short-notice and unplanned coordinated observations with public ToO time. I am sincerely grateful to all involved, as I am to the co-authors of Hynes et al.

(2006), the conclusions of which are summarized here.

REFERENCES

1. Chaty, S., Haswell, C. A., Malzac, J., Hynes, R. I., Shrader, C. R., & Cui, W. 2003, MNRAS, 346, 689
2. Halpern, J. P. 2005, The Astronomer's Telegram, 549
3. Hynes, R. I., Mauche, C. W., Haswell, C. A., Shrader, C. R., Cui, W., & Chaty, S. 2000, ApJ, 539, L37
4. Hynes, R. I., et al. 2003, MNRAS, 345, 292
5. Hynes, R. I. 2005a, ASP Conf. Ser. 330: The Astrophysics of Cataclysmic Variables and Related Objects , 330, 237
6. Hynes, R. I., Gelino, D. M., Pearson, K. J., & Robinson, E. L. 2005, The Astronomer's Telegram, 393
7. Hynes, R. I., & Mullally, F. 2005b, The Astronomer's Telegram, 554
8. Hynes, R. I., & Mullally, F. 2005c, The Astronomer's Telegram, 562
9. Hynes, R. I., Robinson, E. L., Pearson, K. J., Gelino, D. M., Cui, W., Xue, Y., Wood, M. A., Watson, T. K., Winget, D. E., Silver, I. M., 2006, ApJ, submitted
10. Kanbach, G., Straubmeier, C., Spruit, H. C., & Belloni, T. 2001, Nature, 414, 180
11. McClintock, J. E., Garcia, M. R., Caldwell, N., Falco, E. E., Garnavich, P. M., & Zhao, P. 2001, ApJ, 551, L147
12. McClintock, J. E., et al. 2001, ApJ, 555, 477
13. Markoff, S., Falcke, H., & Fender, R. 2001, A&A, 372, L25
14. Morgan, E., Swank, J., Markwardt, C., & Gehrels, N. 2005, The Astronomer's Telegram, 550
15. Morris, D. C., Burrows, D. N., Racusin, J., Roming, P., Chester, M., Verghetta, R. L., Markwardt, C. B., & Barthelmy, S. D. 2005, The Astronomer's Telegram, 552
16. O'Brien, K., Horne, K., Hynes, R. I., Chen, W., Haswell, C. A., & Still, M. D. 2002, MNRAS, 334, 426
17. Palmer, D. M., Barthelmey, S. D., Cummings, J. R., Gehrels, N., Krimm, H. A., Markwardt, C. B., Sakamoto, T., & Tueller, J. 2005, The Astronomer's Telegram, 546
18. Remillard, R., Morgan, E., Smith, D., & Smith, E. 2000, IAU Circ. 7389
19. Still, M., Roming, P., Brocksopp, C., & Markwardt, C. B. 2005, The Astronomer's Telegram, 553
20. Wagner, R. M., Foltz, C. B., Shahbaz, T., Casares, J., Charles, P. A., Starrfield, S. G., & Hewett, P. 2001, ApJ, 556, 42
21. Yuan, F., Cui, W., & Narayan, R. 2005, ApJ, 620, 905
22. Zurita, C., et al. 2005, The Astronomer's Telegram, 383

Infrared Counterparts to X-ray sources in the Galactic Center region

Francisco Jablonski and Leonardo A. Ramos

Instituto Nacional de Pesquisas Espaciais,
Av. dos Astronautas 1758, 12227-010 S.J. dos Campos-SP, Brazil

Abstract. We have carried out a study to identify NIR counterparts to *Chandra* X-ray sources in the Galactic Center region. We used H-band images collected at the Brazilian national facility for astrophysics as well as 2MASS data for this. For the suggested counterparts with observations in more than one occasion, we investigate the likelihood of intrinsic variability.

Keywords: X-ray sources, IR sources, Variable stars
PACS: 98.70.Qy, 98.70.Lt, 97.30.-b

INTRODUCTION

The *Chandra* X-ray telescope has been used by [1] to survey the Galactic Center (GC) region with unprecedented sensitivity and spatial resolution. A catalog of 2357 point sources with L_x as low as 3×10^{33} erg s^{-1} has been produced in a field-of-view of $17' \times 17'$. The positional uncertainty of $\leq 4''$ allows us to look for counterparts of these sources in other wavelengths. In the optical, due to the large extinction by the interstellar medium (tens of magnitudes in the V band), the number of counterparts is small. In the infrared, however, the extinction is relatively low (~ 5 mag at 1.65μm) and many sources like Young Stellar Objects [2][3][4], RS CVn/Algol systems [5][6], Wolf-Rayet binaries [7][8][9], Cataclysmic Variables [10][11][12] and neutron-star/black-hole X-ray Binaries [13][14][15] are bright enough to be detected even at the distance of the GC. We have used the CamIV camera at the National Laboratory for Astrophysics, in Brazil, to search for infrared counterparts to the *Chandra* sources.

OBSERVATIONS AND DATA REDUCTION

The observations were obtained in the H band (1.65μm) using the 0.6-m and 1.6-m telescopes of Laboratório Nacional de Astrofísica, in Brazil, together with the CamIV camera which is based on a HAWAII 1024×1024 pixels Hg-Cd-Te Rockwell Int. array. The image scale is $\sim 0.5''$/pixel for the 0.6-m telescopes and $0.25''$/pixel for the 1.6-m telescope. The field-of-view in one image is $8' \times 8'$ and $4' \times 4'$, respectively. The seeing conditions were always better than $1.5''$, and for most of the cases the exposure time is 5 min, divided in 5 individual exposures of 60 sec. The sky background was measured with the same exposure time in a position separated enough from the GC so that crowding was not an issue for the median combination process. Besides the data collected by us in 2003, we also used archive data publicly available at the National Lab.

The data reduction consists of linearizing the images, subtracting the sky background and flat-fielding the data. An IRAF script that calls the daofind task is used to find the point sources. Only sources relatively free of confusion were considered in our analysis. The conversion from instrumental to standard H band magnitudes is done by means of several isolated objects listed in the 2MASS catalog.

IDENTIFICATION OF NIR COUNTERPARTS

The likelihood of an identifications takes into account the positions of the X-ray and NIR sources and the respective fluxes. Quantitatively,

$$\mathscr{L}_I = \frac{1}{\sigma_X \sqrt{2\pi}} \exp\left(\frac{-|d|^2}{2\sigma_X^2}\right) \times \frac{\sigma_{0H}}{\sigma_H + \sigma_{0H}} \times \frac{\sigma_{0X}}{\frac{\sigma_N}{N} + \sigma_{0X}} \qquad (1)$$

Here d is the difference in position between the X-ray source and the candidate NIR identification, σ_X is the uncertainty (in arcsec) in the X-ray position:

$$\sigma_X = \begin{cases} 0.3 + 0.0234\theta^2 & \theta < 8' \\ 1.5 + 0.875(\theta - 8) & 8' \leq \theta < 12' \end{cases}$$

N is the net counts for the X-ray source so that $\frac{\sigma_N}{N}$ is the fractional error in the X-ray flux and σ_H is the corresponding error in the H band. The factors containing σ_{0X} and σ_{0H} assure that the likelihood is well behaved for both faint and bright sources.

VARIABILITY

The likely NIR counterparts to X-ray sources that were observed in two or more occasions were examined for variability. Here we adopted the approach of [16] modeling the total variance of the measurements as

$$\sigma_{tot}^2 = \sigma_{noise}^2 + \sigma_{intrinsic}^2 \qquad (2)$$

with σ_{noise} being the same as σ_H in Eq. (1). $\sigma_{intrinsic}$ accounts for any other contribution for the variance, including variability due to eruptions, orbital effects, et cetera. Given independent flux measurements y_i, σ_i, $i = 1,...,M$, the likelihood for $\sigma_{intrinsic}$ can be expressed as

$$\mathscr{L}_V(\sigma_{intrinsic}|y_i,\sigma_i) \propto p(y_i|\sigma_i,\sigma_{intrinsic}) = \prod_{i=1}^{M} \frac{e^{-\frac{1}{2}[(y_i-\bar{y})^2/(\sigma_i^2+\sigma_{intrinsic}^2)]}}{4\pi^2(\sigma_i^2+\sigma_{intrinsic}^2)^{\frac{1}{2}}} \qquad (3)$$

In practice, both $\sigma_{intrinsic}$ and the mean flux $\bar{y}$ are calculated iteratively.

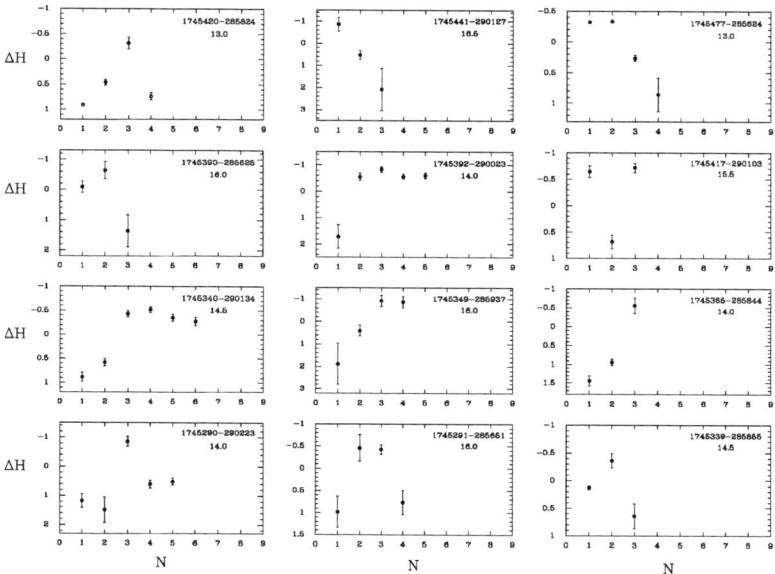

FIGURE 1. A few examples of NIR counterparts found to be variable. The horizontal axis in each panel is just order number of the observation, *not* time. The vertical axis shows the residual magnitudes with respect to the mean magnitude listed below the identification of the X-ray source.

RESULTS

We have found 843 likely counterparts to the *Chandra* sources in the Galactic Center region. From these, 473 sources have been measured in two or more occasions. 55 sources show evidences of variability. Figure 1 shows a few examples of variable sources. A few comments on particular sources:

- 174536.1-285638: The largest X-ray flux. Likelihood of identification is high, but the object is not variable. The X-ray hard color (HR2) is quite "blue", suggesting a foreground object. H=12.25.
- 174537.9-290134: The highest likelihood of identification. Besides being quite bright (H=9.7) it has HR2 slightly negative. Possibly a foreground object.
- 174538.5-290136: The brightest object simultaneously in H (13.1) and X-rays. HR2=0.134. Even considering a relatively small variability ($\sigma_{intrinsic} = 0.4$) this is a good candidate to a CV.
- 174541.7-285823: Brightest (H=11.6) among the variable objects. Relatively bright in X-rays, with HR2=0.1. Could be a foreground, active, late-type star.
- 174534.9-285937: HR2=0.19, and only SFX is low. $\sigma_{intrinsic} = 1.25$. This is a good candidate to a CV.
- 174536.3-285837: Only two measurements, but $\sigma_{intrinsic} \sim 1.6$. Well isolated, with H=16.4, HR2=0.4 and HFX=1.47. It could be a CV or LMXB.

- 174542.0-285824: Variable, with HR2=0.56 and large HFX. With H=13.11 this could also be an X-ray binary.

ACKNOWLEDGMENTS

This work is based on observations made at Observatório Pico dos Dias, operated by Laboratório Nacional de Astrofísica, Brazil. IRAF is distributed by National Optical Astronomy Observatories, which is operated by the Association of Universities for Research in Astronomy, Inc., under contract with NSF.

REFERENCES

1. Muno, M.P., Baganoff, F.K., Bautz, M.W., Brandt, W.N., Broos, P.S., Feigelson, E.D., Garmire, G.P., Morris, M.R., Ricker, G.R., Townsley, L.K., *ApJ* **589**, 225-241 (2003).
2. Garmire, G., Feigelson, E.D., Broos, P., Hillenbrand, L.A., Pravdo, S.H., Townsley, L., Tsuboi, Y., *AJ* **120**, 1426-1435 (2000).
3. Preibisch, T., Zinnecker, H., *AJ* **123**, 1613-1628 (2002).
4. Kohno, M., Koyama, K., Hamaguchi, K., *ApJ* **567**, 423-433 (2002).
5. Singh, K.P., Drake, S.A., White, N.E., *AJ* **112**, 221-229 (1996).
6. Dempsey, R.C., Linsky, J.L., Schmitt, J.M., Fleming, T.A., *ApJ* **413**, 333-338 (1993).
7. Yusef-Zadeh, F., Law, C., Wardle, M., Wang, Q.D., Fruscione, A., Lang, C.C., Cotera, A., *ApJ* **570**, 665-670 (2002).
8. Pollock, A.M.T., *ApJ* **320**, 283-295 (1987).
9. Zwart, S.F.P., Pooley, D., Lewin, W.H.G., *ApJ* **574**, 762-770 (2002).
10. Mukai, K., *NewAR* **44**, 9-13 (2000).
11. Mauche, C.W., Mukai, K., *ApJ* **566**, L33-L36 (2002).
12. Szkody, P., Nishikida, K. Raymond, J.C., Seth, A., Hoard, D.W., Long, K.S., Sion, E.M., *ApJ* **574**, 942-949 (2002).
13. Asai, K. Dotani, T., Hoshi, R., Tanaka, Y. Robinson, C.R., Terada, K., *PASJ* **50**, 611-619 (1998).
14. Wijnands, R., Guainazzi, M., van der Klis, M., Méndez, M., *ApJ* **573**, L45-L49 (2002).
15. Campana, S., Stella, L., Israel, G.L., Moretti, A., Parmar, A.N., Orlandini, M., *ApJ* **580**, 389-393 (2002).
16. Almaini, O., Lawrence, A. Shanks, T., Edge, A., Boyle, B.J., Georgantopoulos, I. Gunn, K.F., Stewart, G.C., Griffiths, R.E., *MNRAS* **315**, 325-336 (2000).

Search for an Infrared Counterpart of IGR J16358−4756

Flavio D'Amico[*], Francisco Jablonski[*], Cláudia Vilega Rodrigues[*], Deonísio Cieslinski[*] and Gabriel Hickel[†]

[*]*Instituto Nacional de Pesquisas Espaciais,*
Av. dos Astronautas 1758, 12227-010 S.J. dos Campos-SP, Brazil
[†]*Universidade do Vale do Paraíba,*
Av. Shishima Hifumi 2911, 12244-000 S.J. dos Campos-SP, Brazil

Abstract. We report here on near infrared observations of the field around IGR J16358−4726. The source belongs to the new class of highly absorbed X-ray binaries discovered by IBIS/*INTEGRAL*. Our primary goal was to identify the infrared counterpart of the source, previously suggested to be a LMXB and then further reclassified as a HMXB. We have made use of *Chandra* observations of the source in order to better constrain the number of possible counterparts. Using the differential photometry technique, in observations spanning a timescale of 1 month, we found no long term variability in our observations. This is compatible, and we suggest here, that the source is a HMXB.

Keywords: Infrared imaging, Near infrared observations, X-ray binaries
PACS: 95.55.Qf, 95.85.Jq, 97.60.-s, 97.80.Jp

INTRODUCTION

Not only, with the advent of the "International Gamma-ray Astrophysics Laboratory", (*INTEGRAL*) [1] mission, has the number of known X-ray binaries increased as well a new category of sources was unveiled by the power of ISGRI/IBIS [2] telescope onboard *INTEGRAL*. The sources belonging to this new category are the so-called highly absorbed hard X-ray sources, with column densities higher than 10^{23} cm^{-2} [3]. The recently discovered (2003 March 19) IGR J16358−4726 source [4] is one within this new category.

After its *INTEGRAL* discovery source was observed by *Chandra* ACIS-2 [5] significantly 9′.7 off-axis imaging spectrometer. *Chandra* has, however, provided a position for the very soft (i.e., $\lesssim$ 10 keV) X-ray counterpart with an 0″.6 error radius [6]. Strong pulsations were also found with a period of ∼ 100 min. After this study the source was classified as being a low mass X-ray binary (LMXB).

Triggered by the *Chandra* study, IBIS/*INTEGRAL* data up to revolution 114 were analyzed [7]. Those authors found the same *Chandra* 1.6 hours flux modulation in the 18–60 keV band. Source was classified, in this work (as usual for the highly absorbed sources), as a high mass X-ray binary (HMXB). Pulsed fraction is high (∼ 70%) both in *Chandra* [5] and *INTEGRAL* [7] data.

Despite of the good quality of the available X-ray data, no counterpart was found yet (infrared, radio, etc.). The identification of infrared counterpart of recently identified *INTEGRAL* sources is the primary goal of an ongoing project by our group. Here we report preliminary results for IGR J16358−4726. In the next few sections we describe the observations, carried out in the near infrared with the available Brazilian instrumentation, then we show both data and data analysis, then our results and finally our conclusions will be presented.

OBSERVATIONS AND DATA ANALYSIS

We observed IGR J16358−4726 for 6 nights in 2004 June 22–24 and July 26–28 at *Laboratório Nacional de Astrofísica* (LNA/MCT, Brazil) using a 1.6 m telescope with the *CamIV* infrared camera (details in [8]). We did H band photometry for sources inside a 5″ circular region centered in *Chandra* position. In our observations seeing conditions were always under 1″, images were taken with 60 s integration time in a field of view of 4′ × 4′. In our nearly 450 images, the magnitude limit is 19.5. Image reduction was done using specific tasks for *CamIV* integrated (and designed) for use within IRAF following standard procedures (like flat-fields, bad pixel mapping, background subtraction and the differential photometry itself). Figure (1) shows a slice 1′.3 × 1′ of the *CamIV* field.

Immediately after the *Chandra* observations an infrared counterpart was suggested based in 2MASS archive [6]. The source is refereed as 2MASS J16355369-4725398, very close to the *Chandra* position at $\alpha = $ 16h 35m 53s.8 and $\delta = -47°25'41''.1$ (J2000). Off-axis images of *Chandra* are known to be inaccurate in terms of centroid positions (see [9] and [10]) by amounts always less than 2″. In this sense our search box is exaggerated.

As usual, since our images are deeper than 2MASS ones, we discovered a new source, located in the very vicinity of the *Chandra* counterpart. We're aware of the danger in doing circular aperture differential photometry rather than point spread function (PSF) extraction for such a source very close to other star, but we decided to take this approach in this preliminary study, and the results have shown that the technique performed quite well.

RESULTS AND DISCUSSION

Our results are shown in Fig. (2), where we display for each image the corrected magnitude and also averaged magnitudes for each of our 6 nights of observation on 2004 (June 22–24, July 26–28).

Possible photometric variability of the sources close to the X-ray error box was quantified in two ways: first, a simple χ^2 test, $\chi^2 = \sum_1^N (y_i - \bar{y})^2 / \sigma_i$, in which the null hypothesis is the constancy of $\bar{y}$. Here $y_1, ..., y_N$ (N=6) are the photometric measurements and $\sigma_1, ..., \sigma_N$ the corresponding errors. Notice that since σ_i are obtained from many differential magnitudes with respect to a well-exposed reference star, these quantities are quite robust, including the contributions of photon noise, scintillation and systematic effects. We obtained $\chi^2 = 0.70$, 2.07, and 3.47 respectively for stars #1, #2, and #3 which corresponds to 98%, 84%, and 63% probability of constancy of the measurements.

The second way of testing for photometric variability is a little more elaborated, following [11]. The variance of the 6 nights of data is modeled as $\sigma_{total}^2 = \sigma_{noise}^2 + \sigma_{intrinsic}^2$, that is, the data are supposed to be characterized by the noise σ_{noise} and

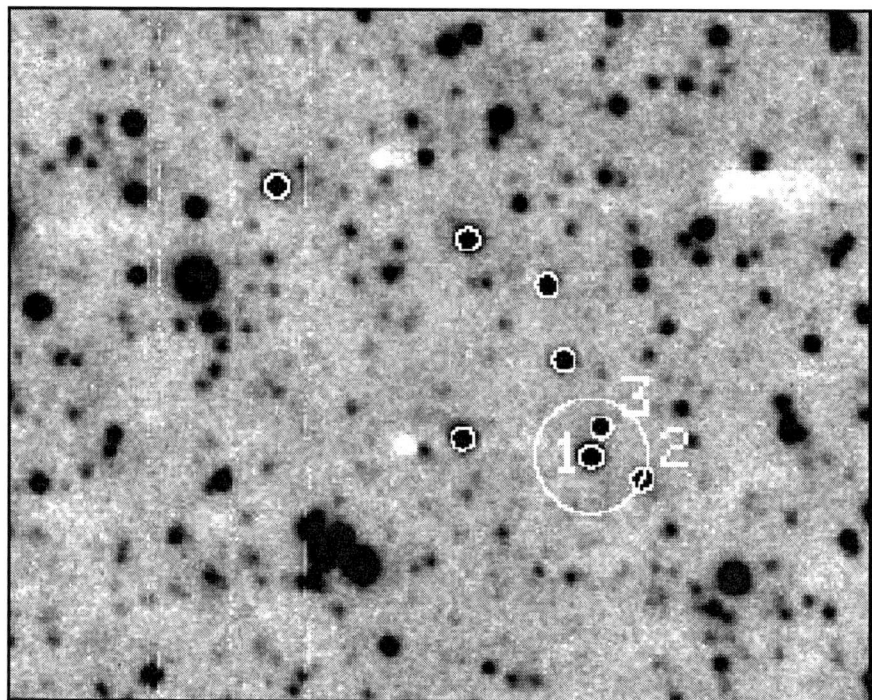

FIGURE 1. A ~ $1' \times 1'$ slice of the entire *CamIV* field including the 2MASS suggested counterpart of IGR J16358−4726 (labeled as 1). We can see clearly the presence of a source in the vicinity of 1 (labeled as 3), not previously known in the 2MASS archive. Around the source #1, our search box of $5''$ is also showed. Other sources used in the differential photometry are marked with small circles. North is on top, East is on right

some intrinsic variations $\sigma_{\text{intrinsic}}$. Maximum likelihood estimates produce $\sigma_{\text{intrinsic}} =$ 0.00, 0.02, and 0.06 mag respectively. These small values, together with the χ^2 results above indicate that the stars do not present any photometric variability, given the data available. It is interesting to note that, if the system is a HMXB, the light of the system must be dominated by the companion, consistent with our results shown here, and confirming the suggestion of [7].

CONCLUSIONS

We have presented here our infrared photometric measurements of IGR J16358−4726 inside a search box of $5''$. The source belongs to a new class of highly photo-absorbed hard X-ray sources. Using a *Chandra* X-ray counterpart, we search for an infrared counterpart to IGR J16358−4726 inside a $5''$ error search box. Upper limits to any variability for objects #1, #2, and #3 are 0.00, 0.02, and 0.06 respectively. If the system is a LMXB, as previously suggested by some authors, then we would expect to see

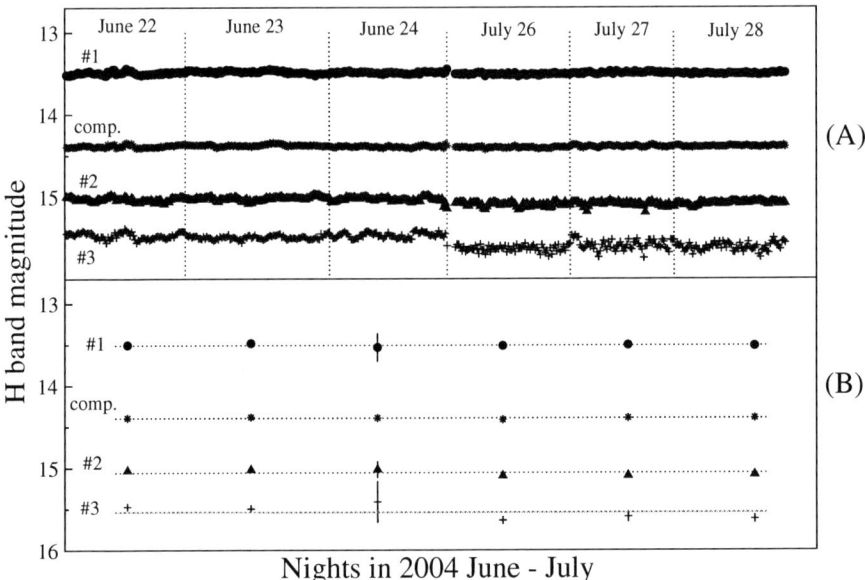

FIGURE 2. (A) The corrected magnitudes for each of the ~ 450 images for the 3 objects (labeled as #1, #2 and #3, as in Fig. 1) inside our search box of $5''$. We also plot the results for the sum of fluxes of several comparison stars (labeled as comp and made dimmer by 2 mags only for plotting reasons). (B) As in (A), but with the computed average magnitude for each night for each object. Horizontal lines are averaged values for the 6 nights. Error bars are also plotted individually (some can't be seen due to its small values).

larger variations, otherwise we are forced to conclude that the infrared counterpart of IGR J16358−4726 is still unidentified. If the system belongs to the class of HMXB, also as previously suggested by some authors, then the light of the system must be dominated by the companion, and no photometric variations are expected: our data, that span a timescale of 1 month, seem to support that interpretation. We suggest that IGR J16358−4726 is a system belonging to the class of HMXB.

ACKNOWLEDGMENTS

This work is based in observations made at Observatório Pico dos Dias, operated by the Laboratório Nacional de Astrofísica, Brazil. FD acknowledges for useful discussions during the Workshop time: thanks a lot!

REFERENCES

1. Winkler, C., et al., *A&A* **411**, L1-L6 (2003).
2. Ubertini, P., et al., *A&A* **411**, L131-L139 (2003).

3. Kuulkers, E., "An Absorbed view of a new class of INTEGRAL sources", in *INTERACTING BINARIES: Accretion, Evolution, and Outcomes*, edited by Luciano Burderi et al., AIP Conference Proceedings 797, New York, 2005, pp. 402-409.
4. Revnivtsev, M., et al., *IAUC 8097* (2003).
5. Patel, S. K., et al., *ApJ* **602**, L45-L48 (2004).
6. Kouveliotou, C., et al., *IAUC 8109* (2003).
7. Lutovinov, A., et al., *A&A* **444**, 821-829 (2005).
8. See details in www.lna.br
9. Giacconi, R., et al., *ApJ Supp. Series* **139**, 369-410 (2002).
10. McHardy, I. M., et al., *MNRAS* **342**, 802-822 (2003).
11. Almaini, O., et al., *MNRAS* **315**, 325-336 (2000).

MAGPIS: The Multi-Array Galactic Plane Imaging Survey

R. H. Becker*, R. L. White[†] and D. J. Helfand**

UC-Davis/LLNL
[†]*Space Telescope Science Institute*
**Columbia University*

Abstract. MAGPIS is a 20 cm radio survey of the first quadrant of the Milky Way. With five-arcsecond angular resolution and 1 mJy sensitivity, it is ideal for detecting faint point sources associated with Galactic transients. The data can be accessed at http://third.ucllnl.org/gps.

Keywords: Radio Surveys; Surveys; Galactic Plane
PACS: 95.55.Jz 95.80.+p

INTRODUCTION

In general, large-area radio surveys have not kept pace with progress at other wavelengths. In part this situation has arisen because most radio telescopes are not capable of producing images with an angular resolution sufficiently high to be free of confusion at high sensitivity and to make unambiguous identifications with faint objects at other wavelengths. The Very Large Array is the only radio telescope that combines the necessary sensitivity and resolution to create surveys comparable to those at optical and infrared wavelengths. The VLA is operated as a four-configuration interferometer; observing in any one configuration imposes a trade-off between angular resolution and sensitivity to structure on large scales. For extragalactic surveys where the vast majority of sources are smaller than 10″, a single VLA configuration will usually suffice, but in the Galactic plane, flux is present at angular scales ranging from arcseconds to degrees, and adequate surveys require multi-configuration imaging making surveying a more expensive proposition. In this paper we report on the first project to use the VLA to carry out an extensive survey of the Milky Way.

A BRIEF HISTORY OF GALACTIC PLANE RADIO SURVEYS

Prior to 1990, our knowledge of radio emission from Milky Way came from surveys made with the 100m Bonn telescope in Effelsberg. This instrument was used to image the Galactic plane at 1.4, 2.7, and 5 GHz, resulting in images with angular resolutions of 9, 4, and 3 arcminutes, respectively (Reich, Reich, & Fuerst 1990; Reich, Fuerst, Reich, & Reif 1990 A&AS; Altenhoff, Downes, Pails, & Schraml 1979 A&AS). These surveys were successful in cataloging the brightest HII regions and supernova remnants (SNRs) in the Galaxy but lacked suffcient angular resolution and sensitivity to identify the majority of such objects. Nonetheless, they have provided the foundation for Galactic radio astronomy for more than a decade.

Starting in early 1990, single-configuration VLA radio surveys of the Galactic plane became available at 1.4 and 5 GHz (Becker et al. 1990; Helfand et al. 1992; Becker et al. 1994). Both had an angular resolution of $\sim 5''$ and a sensitivity of 10 mJy or better. Both surveys resulted in source catalogs containing thousands of compact objects, but they were largely insensitive to more extended structures owing to missing short baselines.

MAGPIS

The shortcomings of both single-dish and single-configuration surveys were readily apparent. It was clear that one could have the best of both worlds by using the VLA in a multi-configuration mode, and then adding the single-dish maps to fill in the largest angular scales. In 2000, we began the Multi-Array Galactic Plane imaging Survey (MAGPIS) program, collecting 1.4 GHz VLA observations in three configurations (B, C, & D). The initial phase of the survey covered the longitude range $19° < l < 32°$. Two years later, phase II followed, covering $5° < l < 19°$; in late 2005, phase III initiated coverage up to $65°$ to complete the northern region imaged by the Spitzer Legacy program GLIMPSE (Benjamin et al. 2003). All three phases cover a latitude range $|b| < 0.8°$. The data from the first two phases are fully reduced and online at http://third.ucllnl.org/gps; the third phase should become available towards the beginning of 2007.

The MAGPIS images have $\sim 5''$ angular resolution and a point source sensitivity over most of the images of $\sim 1-2$ mJy. The MAGPIS website provides access to five sets of images: the MAGPIS 1.4 GHz survey, the older single-configuration images at 1.4 and 5 GHz, a 327 MHz image from the VLA, and 21μm images from the MSX satellite (Price et al. 2001). The latter images are particularly useful for distinguishing thermal and nonthermal Galactic sources (i.e., separating HII regions from SNRs). In general, thermal sources are both radio and mid-IR emitters, while SNRs are strong in the radio and weak in the mid-IR. Examples of how radio and mid-IR data can be used to discriminate between thermal and nonthermal emission are given in the MAGPIS introductory paper Helfand et al. (2006).

TRANSIENT RADIO SOURCES IN THE MILKY WAY

The MAGPIS website can be used to search for transient sources by comparing the various surveys presented there. In particular, comparison between the archival snapshot 1.4 GHz survey from the early 1990s with the newer MAPGPIS survey can reveal extreme examples (extreme because the earlier survey is of much lower quality and hence only gross examples of variability should be trusted). In addition, since MAGPIS is collected in three different arrays, each area of the survey is visted three times at intervals of 4 to 12 months. As a consequence of the very different resolutions afforded by each visit, evidence for variability from the multiple visits must also be treated with caution. Furthermore, the MAGPIS website does not provide separate images for each visit so one must go to the original data for such an inspection.

An example of the use of MAGPIS to study variability of Galactic plane sources is provided by the discovery of the first radio emission from a magnetar. The transient anomalous X-ray pulsar XTE J1810-197 was found to be a 4.5 mJy source in the MAGPIS images (Halpern et al. 2005). Analysis of the individual configuration data showed it clearly present in B-array observations taken in January 2004, one year after its X-ray outburst; all earlier observations yielded upper limits roughly consistent with, or slightly below, this value; followup observations are planned.

We have yet to conduct a systematic survey for variability either within the MAGPIS data or by comparing these images to earlier archival data. The discovery of the radio counterpart to XTE J1810-197, along with earlier work on Galactic plane variability (Gregory & Taylor 1986; Freismuth, Langston, & Minter 2002), suggests such efforts would be worthwhile.

ACKNOWLEDGMENTS

The MAGPIS survey is supported in part by the National Science Foundation under grants AST-05-07598 and AST-05-07663.

REFERENCES

Altenhoff, W. J., Downes, D., Pauls, T., & Schraml, J. 1979, A&AS, 35, 23.
Becker, R. H., White, R. L., McLean, B. J., Helfand, D. J., & Zoonematkermani, S. 1990, APJ, 358, 485.
Becker, R. H., White, R. L., Helfand, D. J., & Zoonematkermani, S. 1994, APJS, 91, 347.
Benjamin, R. A., et al. 2003, PASP, 115, 953.
Freismuth, T. M., Langston, G. I., & Minter, A. H. 2002, Bulletin of the American Astronomical Society, 34, 716.
Gregory, P. C., & Taylor, A. R. 1986, AJ, 92, 371.
Halpern, J. P., Gotthelf, E. V., Becker, R. H., Helfand, D. J., & White, R. L. 2005, APJL, 632, L29.
Helfand, D. J., Zoonematkermani, S., Becker, R. H., & White, R. L. 1992, APJS, 80, 211.
Helfand, D. J., Becker, R. H., White, R. L., Fallon, A. and Tuttle, S. 2006 APJ, in press (astro-ph 0510468).
Price, S. D., Egan, M. P., Carey, S. J., Mizuno, D. R., & Kuchar, T. A. 2001, AJ, 121, 2819.
Reich, W., Reich, P., & Fuerst, E. 1990, A&AS, 83, 539.
Reich, W., Fuerst, E., Reich, P., & Reif, K. 1990, A&AS, 85, 633.

MIRAX INSTRUMENTS AND SOFTWARE

CZT Detector and HXI Development at CASS/UCSD

Richard E. Rothschild, John A. Tomsick, James L. Matteson,
Michael R. Pelling, and Slawomir Suchy

Center for Astrophysics and Space Sciences
University of California, San Diego
La Jolla, CA 92093-0424
USA

Abstract. The scientific goals and concept design of the Hard X-ray Imager (HXI) for *MIRAX* are presented to set the context for a discussion of the status of the HXI development. Emphasis is placed upon the RENA ASIC performance, the detector module upgrades, and a planned high altitude balloon flight to validate the HXI design and performance in a near-space environment.

Keywords: Astronomical Instrumentation, X-ray and Gamma-ray, Imaging Detectors, Spectroscopy
PACS: 95.55.Ka, 95.75.Fg, 95.85.Nv

HXI SCIENTIFIC GOALS

The HXI scientific goals apply to the entire *MIRAX* mission, and are two-fold: 1) To be highly efficient at discovering short-lived, hard X-ray phenomena, as well as transients on all timescales, and 2) To extend studies of hard X-ray recurrent transient and persistent sources. HXI sensitivity over the <10 keV to 200 keV band will allow for discovery of transients brighter than 10 mCrab in an hour. HXI localization accuracy will provide positions accurate to one arcminute for a 10σ detection, and this in turn can lead to identification of the object at other wavelengths. The continuous nature of the *MIRAX* observing of the central region of the Galactic plane allows for monitoring of the transient activity throughout the outburst, no matter how long or short it may be. Figure 1 Left shows the HXI sensitivity limit as a function of outburst duration. HXI clearly carves out a unique region of transient discovery space with respect to present hard X-ray missions that feature Galactic monitoring programs.

Several tens of persistent hard X-ray sources will be in the 10^3 square degree field of view and this allow *MIRAX* to extend the detailed studies of these objects, first begun by earlier missions, such as the *Ariel V* and *RXTE* All-Sky Monitors. From these observations, HXI will observe state changes and other examples of aperiodic variability, and will generate a hard X-ray Galactic bulge catalog of sources.

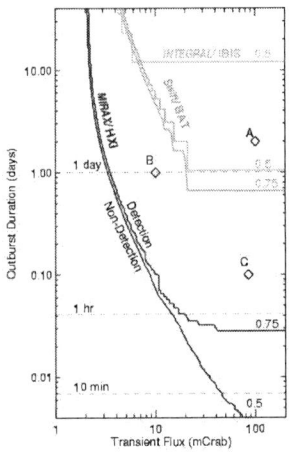 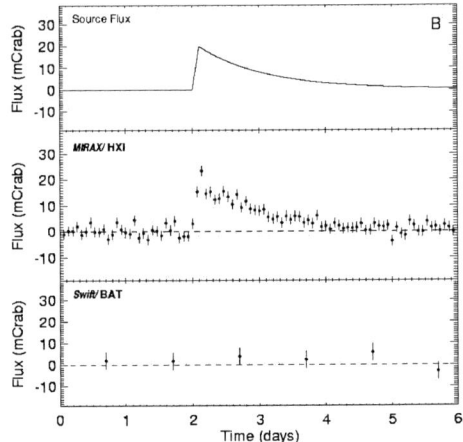

FIGURE 1. Left: The HXI transient flux sensitivity expressed as a function of outburst duration. This is compared to that for the *Swift*/BAT and *INTEGRAL*/IBIS instruments. The two sets of lines demark the regions of detection and non-detection at the 50% and 75% probability level for each instrument. **Right:** Comparison of HXI with Swift for a 20 mCrab transient that is above 10 mCrab for a day with a fast rise and exponential decay. The short duration precludes INTEGRAL from having a high probability of detection. This example is denoted B and its position on the sensitivity plot (**Left**) is shown as a diamond below the letter B.

TABLE 1. Comparison of Monitoring Coverage with Various Missions.

Single Observation: Mission/Instrument	Frequency (per day)	Duration (hours)	Sensitivity (mCrab)	Systematic Limit (mCrab)	Annual Coverage
MIRAX/HXI	15	1	10	2	75%
Swift/BAT	1	0.2	20	3	75%
INTEGRAL/ISGRI(GPS)	1/12	0.8	5	2	<50%

The power of *MIRAX* and the HXI can be depicted in two ways. First of all, Table 1 compares the monitoring coverage of HXI with that of the two other active Galactic plane monitoring missions. *Swift* is primarily a gamma-ray burst mission whose operations will provide for observing the available sky each day, and *INTEGRAL* has a Galactic Plane Survey (GPS) and Galactic Center Deep Exposure (GCDE) pointing programs that are subject to strong solar constraints. These programs will have only 1% (Swift), 0.2% (GPS), and 12% (GCDE) detection probabilities for rapid transients independent of source flux. This is to be compared to 50% for *MIRAX*.

In terms of making detailed studies of transients, Fig. 1 Right shows a simulated light curve for a 20 mCrab peak flux transient that is greater than 10 mCrab for a day. This illustrates the fine detail that *MIRAX* will obtain in contrast to the sparse sampling to be expected from *Swift*. *INTEGRAL*/ISGRI will only be able to catch longer transient events and then only with infrequent sampling. The qualitative difference is enormous and will lead to physical insights possible only with instruments dedicated to this observing strategy.

HXI CONCEPT DESIGN

The HXI consists of two identical imagers consisting of a tungsten coded mask supported in front of a ~340 cm^2 CZT detector array composed of 9 identical CZT detector modules (Fig. 2, Left). The modules each contain four 32x32x2 mm^3 detectors with 0.5 mm pitch crossed strip electrodes in a 2x2 array for a total detector area of 41 cm^2 (Fig. 2, Right). The anode and cathode strips of the four detectors within a module are electrically interconnected to form an effective 128x128 pixel sensor. Analog and digital support electronics reside directly behind the detector plane, making the overall packaging very compact. The imager support structure holds the coded mask, a graded-Z passive shield, and a 5-sided plastic scintillator active shield around the detector sides and rear to reject prompt charged particle induced background.

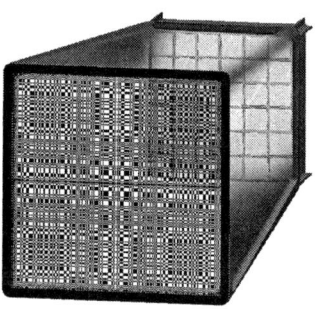 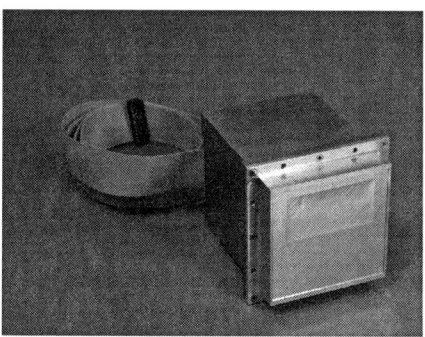

FIGURE 2. Left: The HXI imager with coded mask and detector plane shown. **Right:** The prototype HEXIS module viewed from the incident X-ray direction. All electronics are located below the detectors in the ruggedized housing.

The 256 electrodes are readout by 8 custom ASICs, called RENAs, which provide front end amplifiers and peak detection and hold for each channel. Trigger signals determine which channels are pulse height analyzed and sent to the flight computer for determination of X/Y position, depth of the interaction within the detector, total energy, and event time for each incident photon. Each RENA contains 32 separate amplifier signal chains which are controlled by a gate array logic system. A single HXI detector array then represents a 384x384 grid of 0.5x0.5 mm^2 pixels with which to measure the shadow cast by the coded mask due to the pattern of X-ray sources being observed.

HXI MODULE STATUS

The HXI module work in recent months has centered upon developing a method for CZT detector acceptance testing and improving the RENA ASIC performance.

CZT Detector Selection/Acceptance Testing

Acceptance testing of CZT patterned detector wafers received from the manufacturer is based upon analysis using the UCSD CZT Detector Test/Acceptance Station, which was developed for just this situation. The Station is shown in Fig. 3, and is comprised of the test station, support electronics, and a desktop computer.

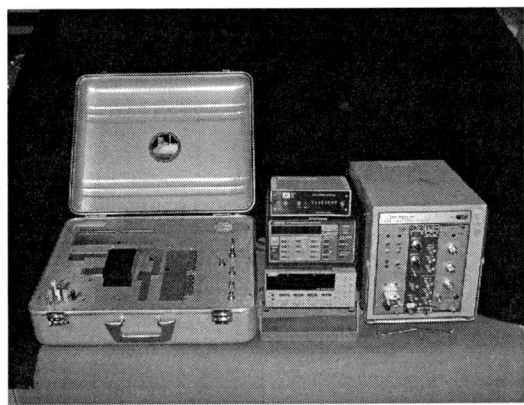

FIGURE 3. The UCSD CZT Detector Test/Acceptance Station with electronics including a low and high voltage supply, Keithley electrometer, and NIM logic units. The latter are used in conjunction with discrete lab electronics contained within the station.

Within the "suitcase" is a complete set of RENA-based electronics, as well as a set of discrete AmpTek amplifiers that can be used for validating RENA results. A complete set of switches, one for each electrode, allow for selection of subsets of the electrodes for test. When used in conjunction with the Keithley switching unit, the resistivity measurements can be completely automated. Acceptance criteria and measurement procedures have been developed to assure CZT detectors meet all the HXI performance specifications.

RENA Development Status

Early testing of the RENA ASIC revealed several changes that needed to be made. Despite the need for the changes, the RENA was incorporated into the initial HEXIS module (Fig. 2, Right) and performance tested. Measurements at 22, 60 and 122 keV demonstrated an energy resolution of 4-5 keV and a sub-10 keV low energy threshold. A total redesign of the chip has been accomplished and the chip has been under test at both UCSD and Nova, Inc. The principal changes of 1) electrode level threshold/polarity settings as opposed to chip level settings, 2) AC-coupled amplifier inputs, and 3) sub-microsecond timing of events (not implemented for HXI) have been validated through initial testing. Early tests with a small pixel detector imply a noise level of <200 electrons rms for the ASIC. When attached to a HXI CZT detector, an energy resolution of <2 keV at 22 keV resulted (see Fig. 4).

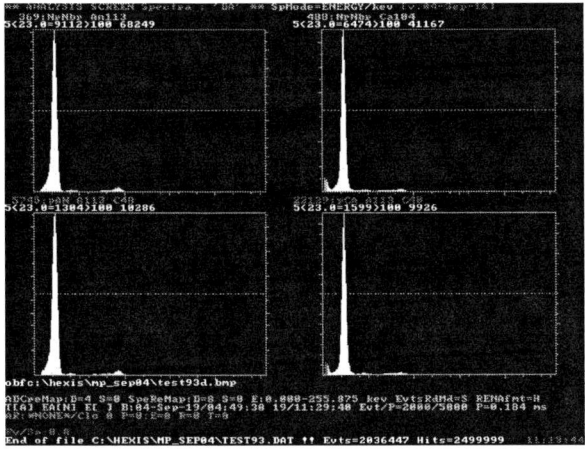

FIGURE 4. Results from testing with the improved RENA ASIC. The upper plots show the pulse height spectra accumulated over an entire anode and cathode strip, respectively, while the lower plots show the anode and cathode spectra for a single pixel, defined as events for which the specified anode and cathode triggered simultaneously.

New electronics boards that are compatible with the existing mechanical housings and previous CZT readout boards have been designed for the improved RENA design, and are in manufacture. This redesign was accomplished within the footprint of the old design, thereby not necessitating any changes to housings or thermal designs.

HEXIS BALLOON FLIGHT

UCSD and Tübingen are preparing a balloon instrument and supporting gondola system utilizing a coded mask and detector based upon the HXI module, and will observe one or more bright Galactic X-ray sources on its initial flight scheduled for late 2006. This is intended to validate the HXI design via spectral and imaging results, and provide background measurements for the development of a detailed background model for HXI. The first flight will have a single HXI module, and subsequent flights will expand the detector plane to four equivalent modules in a 2x2 array. The latter configuration will validate the multi-module design.

Included in the balloon instrument will be a novel particle anticoincidence subsystem surrounding the module and extending to the coded mask. This plastic scintillator array will be comprised of 13 segments – a 5 sided box directly surrounding the module and two sets of 4 elements forming a square tube extending to the mask. All 13 segments will utilize wavelength shifting fibers to collect the scintillation light and direct the shifted green light to a 16 element multi-anode PMT. This will allow for analysis of the contribution of each segment of the shield to the overall rejection of charged particles. This, in turn, will allow for design of an efficient HXI antiparticle shield.

HXI Imaging Simulations and Sensitivity

Jorge Mejia and João Braga

Instituto Nacional de Pesquisas Espaciais (INPE), Av. dos Astronautas 1758, São José dos Campos, SP, 12227–010, Brazil.

Abstract. The Hard X-ray Imagers (HXI) of the MIRAX satellite mission are two coded-mask cameras whose main objective is to obtain deep images and spectra (10 to 200 keV) of the central Galactic Plane source population. In this work, we present the results of Monte Carlo simulations of the instrumental background in orbit and the imaging capabilities of the HXIs. These simulations were carried out using the GEANT-based MGGPOD suite of routines for the study of interaction of high-energy particles and photons with the materials of the detector system and its surroundings.

Keywords: MIRAX, Galactic Center, Coded mask, Background, MGGPOD.
PACS: 95.55.Ka, 95.75.Mn, 95.85.Pw, 95.85.Nv

INTRODUCTION

MIRAX (after Monitor e Imageador de RAios-X, in Portuguese) is a satellite mission being developed in Brazil, with strong international partnership, for long-term monitoring of the central Galactic Plane region in the hard and soft X-ray energy bands. For high-energy observations, MIRAX will include two identical coded-mask cameras with a pointing axis offset of 29°, so as to cover a 58° FWHM field of view along the Galactic Plane. MIRAX will be launched on a ~550 km-height circular, equatorial orbit, and will point in the general direction of the Galactic Center during ~9 months per year.

In order to get a better understanding of the instrumental background induced by the space environment, we carried out Monte Carlo simulations of the particle and photon field interactions with the mass model of the HXIs, using the MGGPOD suite developed by Weidenspointner et al. (2004, 2005). This code is a set of Monte Carlo packages and is intended to simulate both prompt and delayed instrumental backgrounds, including the energy deposition of cosmic-ray particles as well as the decay of radioactive isotopes produced in nuclear interactions. Additionally, we have used MGGPOD to study the response of the HXIs to a set of astrophysical sources.

The HXIs will use a 139×139 MURA (Gottesman and Fenimore 1989) coded mask, extended to 242×211 elements, located 600 mm away from the detector plane. The coded mask elements will be made of tungsten with dimensions 1.3 mm × 1.3 mm × 0.5 mm. The detector plane will be composed by a set 3×3 crossed-strip detector modules, each one being a 2×2 array of 32 mm × 32 mm × 2 mm CZT detectors. Each camera will be surrounded by an active plastic scintillator shield and a passive Pb-Sn-Cu graded shield. More details on the HXIs and the MIRAX satellite can be found in

Braga et al. (2004). The HXI mass model, as used in the simulations, is shown in Fig. 1.

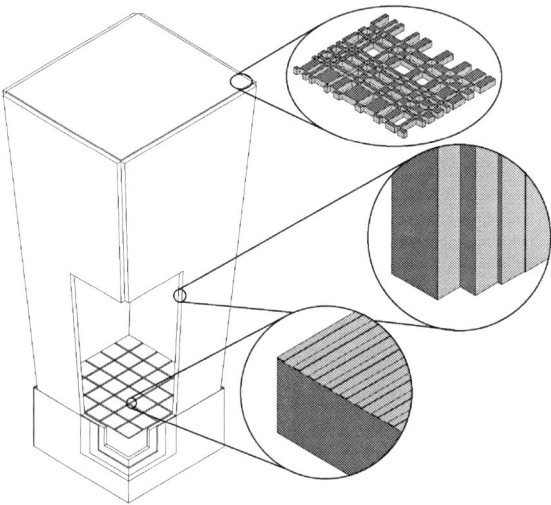

FIGURE 1. HXI mass model. In expanded view, from top to bottom: coded mask, lateral shielding and detector plane.

DETECTOR BACKGROUND MODELLING

Environment Particle Spectra and Flux

Considering the altitude of MIRAX's orbit, particle spectra and fluxes were obtained from the SIREST project home page (http://sirest.larc.nasa.gov) to model the Galactic cosmic-ray proton and SAA trapped proton contributions to be used as input for MGGPOD. In our case, data from SIREST correspond to a 570-km height, 5° inclination circular orbit for the 2-year long period between 01/01/2009 and 12/31/2010. Resulting spectra are shown in Fig. 2. Notice that SIREST spectra are provided for specific epochs and corrected for geomagnetic cutoff. Even though SAA trapped protons contribute to the background only during a fraction of the orbit, they were considered in the determination of the prompt background as if always present.

Additionally, we have considered an isotropic diffuse gamma-ray background with a power-law spectrum of the form (Lei 1997)

$$F_{gamma} = 0.014 \times E^{-2.3} \, photons \, cm^{-2} s^{-1} MeV^{-1} sr^{-1}$$
$$= 111.2 \times E^{-2.3} \, photons \, cm^{-2} s^{-1} keV^{-1} sr^{-1}. \quad (1)$$

Only photons with energies between 20 keV and 10 MeV were considered.

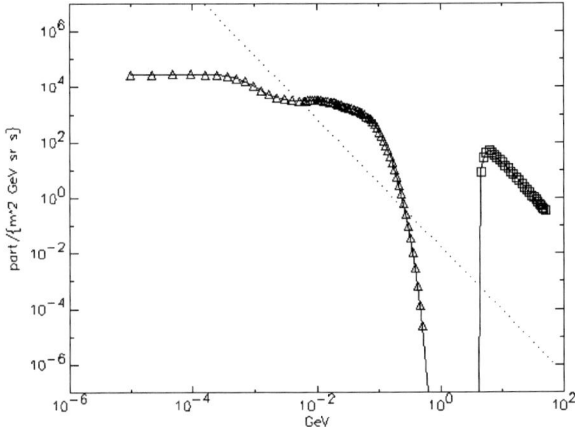

FIGURE 2. Galactic cosmic-ray proton (squares) and SAA trapped proton (triangles) spectra for a MIRAX-like orbit, provided by SIREST, as well as the diffuse photon background spectrum (dots).

Background Results on HXI

Galactic cosmic rays incident on our current HXI mass model produce a prompt background flux of ~8.1 events/s in the detector of each camera. The background flux is reduced to ~0.8 events/s when energy depositions outside of the 20-200 keV band and interactions in more than one element of the camera are removed. During passage near the SAA, the flux slightly increases by an amount of ~0.1 events/s, again, not considering multiple coincidences and limiting the energy band (those values were obtained simulating exposure times of 3600 seconds). Including the delayed interactions, produced by activation and radioactive decay of the camera's materials, additional ~0.69 events/s are produced. Diffuse gamma-rays produce ~33 events/s, which is reduced to ~21 events/s if multiple interaction events are not considered, as well as those corresponding to energy deposits out of the 20 to 200 keV band. Many lines are produced as the result of those interactions. Individual and combined spectra corresponding to each of the aforementioned components are shown in Fig. 3. Using these values, we can calculate the sensitivity in different energy bands and produce the curve shown in Fig. 4. The minimum detectable fluxes are given by the usual expression

$$F_k = \frac{k}{\eta \varepsilon f(1-t)} \sqrt{\frac{B_i + \varepsilon[f + t(1-f)]B_d \frac{G}{A}}{\Delta E A T}}. \quad (2)$$

The expected sensitivity for the entire 20-200 keV energy range is 2.2 mCrab, for an exposure time of 24 hours long exposure time and a 5 σ detection threshold.

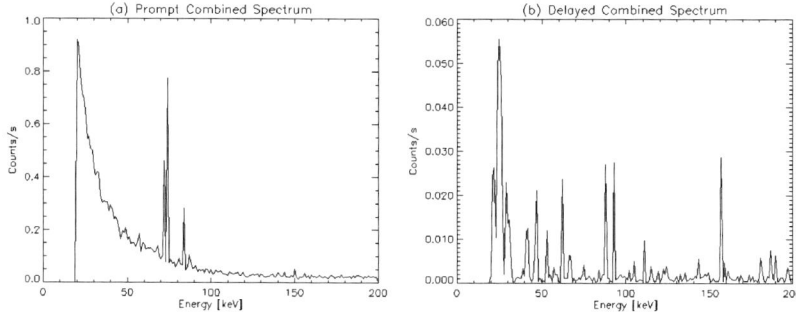

FIGURE 3. Spectra of the simulated prompt (a) and delayed (b) backgrounds produced in the HXI's detectors.

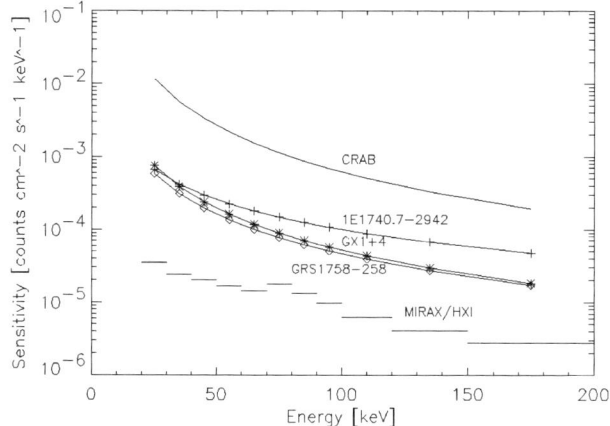

FIGURE 4. Sensitivity curve for the HXIs.

IMAGING WITH THE HXI

In order to test the imaging capabilities of the proposed HXI configuration, we have used MGGPOD to simulate the interaction of photons from a set of astrophysical sources, whose parameters are indicated in Table 1. The reconstruction was carried out using the cross-correlation between the shadowgram cast on the detector and the reconstruction array (Fenimore and Cannon 1981), performed in Fourier space, and is shown in Fig. 5. When the image reconstruction included the PCFOV, the reconstruction array was extended and padded with zeros outside the central 139×139-element region, to guarantee that no side lobes are produced in the corresponding area of the image (Goldwurm et al. 2003). It can be seen that the discrete nature of the detector introduces a cross-like pattern on the reconstructed image because of the lost of coding information in the gaps between CZT elements. This effect was softened by filling in the gaps with values drawn from a pseudo-random distribution with mean and standard deviation equal to that of the shadowgram.

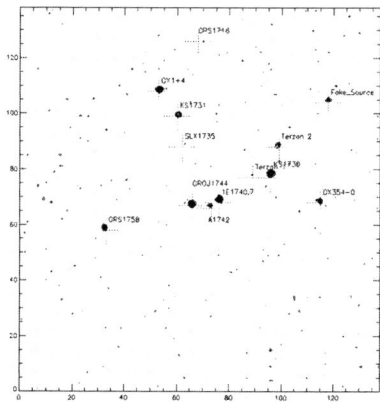

FIGURE 5. Galactic Center region simulated image, including a subset of the sources expected to be detected by HXI, reconstructed over an area of 139×139 sky bins. Exposure time = 24 h. Significance contours begin at 4 σ and increase every 2 σ. Galactic longitude (X-axis) increases to the left.

TABLE 1. List of GC sources included in the image simulation

Source	Flux$_{max}$ [mCrab] (40-100 keV) (Revnivset et al. 2004)	Flux$_{max}$ [photons cm^{-2} s^{-1}] (40-100 keV)	Coordinates (ℓ, b)
1E1740.7-2942	150	0.01452	(-0.87, -0.10)
GRS1758-258	100	0.00968	(4.51, -1.36)
GX1+4	100	0.00968	(1.95, 4.80)
GX354-0	100	0.00968	(-5.69, -0.15)
GX339-4	430	0.041624	(-21.06, -4.32)
SLX1735-269	20	0.001936	(0.81, 2.41)
KS1731-260	70	0.006776	(1..07, 3.66)
Terzan 2	50	0.00484	(-3.68, 2.33)
Terzan 1	10	0.000968	(-2.45, 0.99)
A1742-294	30	0.002904	(-0.45, -0.35)
GROJ1744-28	120	0.011616	(0.39, -0.30)
GRS1716-249	1200	0.11616	(0.15, 6.99)
KS1730-312	170	0.016456	(-3.35, 1.06)

ACKNOWLEDGMENTS

J. Mejia is supported by CNPq grant 381985/2004-0.

REFERENCES

1. G. Weidenspointner, M. J. Harris, C. Ferguson et al., *New Astr. Rev.* **48**, 227-230 (2004).
2. G. Weidenspointner, M. J. Harris, S. Sturner et al., *APJSS* **156**, 69-91 (2005).
3. S.R. Gottesman and E.E. Fenimore, *App. Opt.* **28**, 4344-4352 (1989).
4. J. Braga, R. Rothschild, J. Heise et al., *Adv. Spa. Res.* **34**, 2657-2661 (2004).
5. F. Lei, *The Integral Mass Model*, IN-MM-SOT-TN-0005 (1997).
6. E.E. Fenimore and T.M. Cannon, *App. Opt.* **20**, 1858-1864 (1981).
7. A. Goldwurm, P. David, L. Foschini et al., *A&A* **411**, L223-229 (2003).
8. M.G. Revnivtsev, R.A. Sunyaev, M.R. Gilfanov et al., *Astr. Lett.* **30**, 527-533 (2004).

Event Pre Processor for the CZT Detector on MIRAX

Eckhard Kendziorra[*], Thomas Schanz[*], Slawomir Suchy[¶], and Giuseppe Distratis[*]

[*]*Institut für Astronomie und Astrophysik der Universität Tübingen, Abteilung Astronomie, Sand 1, 72076 Tübingen, Germany*
[¶]*Center for Astrophysics and Space Sciences, University of California, San Diego, La Jolla, CA 92093-0424, USA*

Abstract. We describe the Event Pre Processor (EPP) for the Hard X-ray Imager (HXI) on MIRAX. The EPP provides on board data reduction and event filtering for the HXI Cadmium Zinc Telluride strip detector. Emphasis is placed upon the EPP requirements, its implementation as VHDL design in a Field Programmable Gate Array (FPGA), and the description of a test environment for both the VHDL code and the FPGA hardware.

Keywords: Astronomical Instrumentation, X-ray and Gamma-ray, Imaging Detectors
PACS: 95.55.Ka, 95.85.Nv, 95.75.-z

INTRODUCTION

The scientific objective of the MIRAX satellite mission [1] is the continuous observation of the central 1000 square degrees of the Galactic plane in the energy band 2 – 200 keV. This will allow for the first time to observe light curves of persistent sources on time scales from sub-seconds to months, as well as the discovery of short-lived X-ray events with high sensitivity over a large band pass.

MIRAX is equipped with one Soft X-ray Imager (SXI) [2] and two identical Hard X-ray Imagers (HXI) [3]. Both instruments are position sensitive X-ray detectors behind a coded aperture mask. Currently two options for the data and command interface of the instruments with the On Board Computer (OBC) are discussed: 1) Data from the three detectors are collected and further processed by the Central Electronics Unit (CEU) before transmission to the OBC (see Fig. 1, left). 2) Only HXI data are processed by the CEU, the data and command interface to the SXI is handled directly by the OBC (see Fig. 1, right).

In this paper we discuss the on board data processing of HXI events and describe the current status of the electronics design.

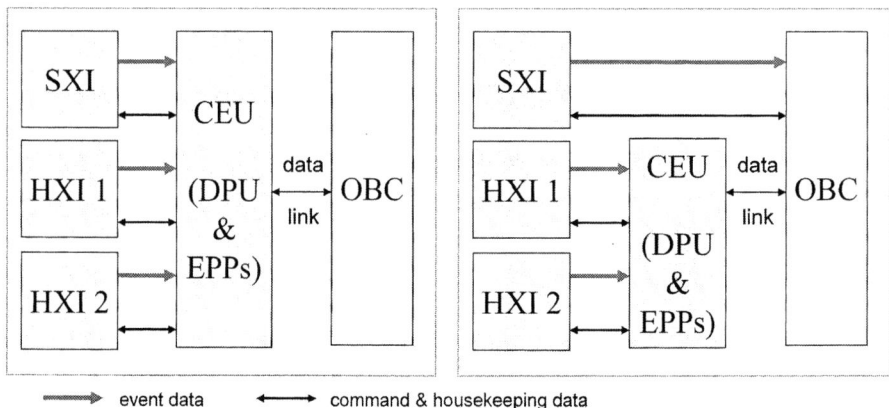

FIGURE 1. Two options, currently studied during phase A, for the interconnection of the instruments with the S/C On Board Computer (OBC)

EVENT PRE PROCESSOR (EPP) FOR HXI

Each of the two Hard X-ray Imagers consists of an array of nine identical detector modules, each containing four 32 x 32 x 2 mm^3 CdZnTe strip detectors with 64 anodes and 64 cathodes perpendicular to each other on both sides of the detector. Figure 2 shows a schematic view of the detector array with its 384 x 384 pixels. The electrodes are read out by custom ASICs, called RENAs, which are placed close to the electrodes. For a charge packet detected on an electrode a RENA chip delivers 14 bit amplitude, 16 bit time stamp, 7 bit channel number and three data status flags. This data package from one electrode is further called a hit.

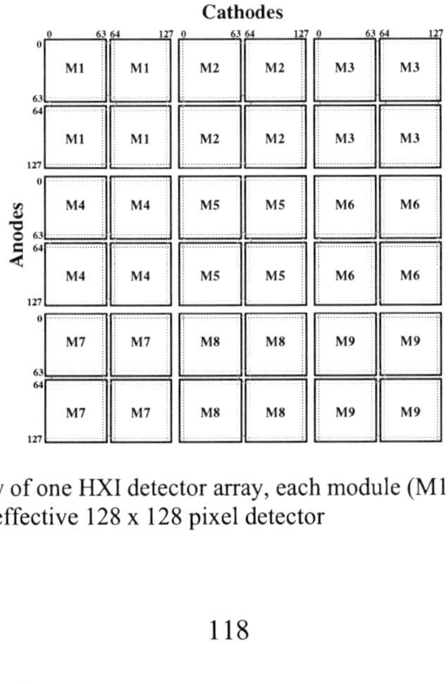

FIGURE 2. Schematic view of one HXI detector array, each module (M1 – M9) contains four CZT strip detectors providing an effective 128 x 128 pixel detector

An X-ray photon interacting in the CZT detector will trigger up to three anode and three cathode hits. These hits have to be further processed and combined on board by an event pre processor (EPP) to a common event data packet. In detail the EPP has to:

- correct the non linearity of the preamplifier, i.e. perform for each anode and each cathode hit a second order polynomial amplitude correction $E_{out} = A + B \cdot E_{in} + C \cdot E_{in}^2$,
- determine the coordinate X of the event as the coordinate of the cathode with the highest amplitude,
- determine the coordinate Y of the event as the coordinate of the anode with the highest amplitude,
- add up the corrected amplitudes of all anodes A_i and all cathodes C_i $A = \Sigma A_i$ and $C = \Sigma C_i$,
- determine the depth of interaction as the ratio between A and C,
- expand the time stamp of hits from the same event to an event time stamp,
- select valid events within an energy band for transmission to telemetry.

The necessary corrections described above are discussed in more detail by Suchy et al. [4]. Two options for the implementation of the EPP tasks will be studied during phase A, a software solution and a dedicated hardware EPP prototype, which has already been developed at IAAT [5]. The design of this hardware EPP is further discussed in the next chapter.

Hardware EPP

A functional diagram of the EPP design utilizing the hardware description language VHDL is shown in Fig. 3. The Control Interface (CTI) connects the EPP to the front end detector electronics, it configurates the RENAs, and receives hits from the RENA board. Only valid hits stemming from the same event are further reformatted in the Event Packet Assembler (EPA) and sent as a 168 bit wide data packet to the Energy Correction (EC) unit. Here, a second order polynomial amplitude correction of up to three anode and three cathode amplitudes is performed in parallel. The eight bit values for the three coefficients (A, B, C) per electrode are stored as a 768 x 24 bit wide reprogrammable look-up table (LUT) in a 1024 x 24 bit RAM. The EC unit also assigns the final X and Y coordinates to the event. Further the depth of interaction (DOI) is computed as the ratio of anode and cathode amplitudes and rounded to a 3 bit value. This is accurate enough to further correct the energy of an event for charge losses of the electron cloud during its drift to the anodes. The corrected event package is then sent to the Filter & Decision (FD) unit for final evaluation. Valid data are defined as follows:
- the event energy is above a lower threshold and below an upper threshold,
- no error had occurred during processing (i.e. no range overflow during amplitude correction, no LUT address error, no division by zero, no signal loss, no signal ratio error),
- no anticoincidence and no multi-site event (i.e. > 3 hits from anodes or cathodes).

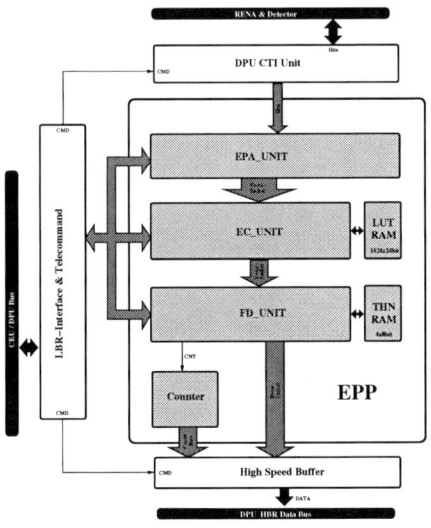

FIGURE 3. EPP design including units and peripheral components

The FD unit writes the final data word (19 bit time stamp, 9 bit X, 9 bit Y, 8 bit energy, and 3 bit DOI) into a FIFO which is read by the data processing unit (DPU). All valid event packets sent to the DPU are counted. This allows us to quantify data losses which might occur during transmission to the ground station

Test Bench

A test environment, called test bench, has been developed at IAAT to verify the functionality of the EPP. Here, the complete EPP design or a subset can be loaded into

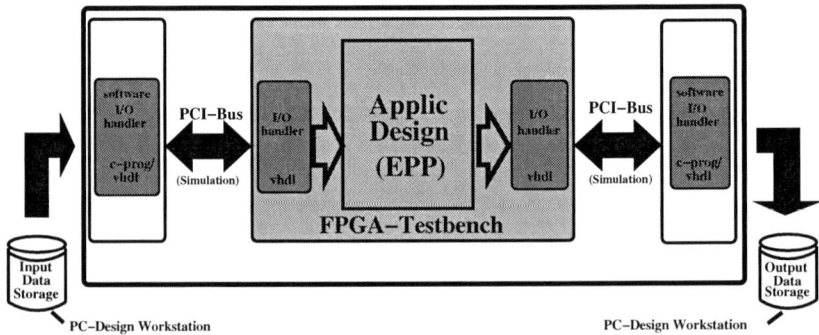

FIGURE 4. Concept of the EPP test bench, the test bench consists of two parts: 1) The FPGA test bench inside the XILINX FPGA board enclosing the application design and the VHDL handler. 2) The I/O software and the VHDL code which is necessary for data transfer between the hard disk and the PCI Bus.

a XILINX FPGA prototype evaluation board installed as a PCI card in a common PC-type computer. We have also designed a purely software based VHDL simulation environment in order to verify the results of the FPGA test bench measurements. Both, the VHDL simulator environment and the hardware test bench use the same input files. Because the XILINX board can only communicate with the host computer via the PCI Bus and due to the fact that neither the synthesized EPP design nor the FPGA test bench can control hard disk operations, there is a need for a data handling software. This software enables a data transfer with 32 bit words between the hard disk and the FPGA board via the PCI Bus. The test bench provides counterpart VHDL based I/O handlers to convert the data width between the 32 bit PCI bus and the EPP I/O event data packet format. A communication between I/O software and VHDL I/O handler has to be assured all the time. Figure 4 shows the concept and data flow of the EPP test bench.

The simulation with VHDL reduces the development time significantly and enables functional tests without time consuming synthesis of the design. The software I/O program is replaced by VHDL code to handle hard-disk I/O during simulation. Hence the design can be verified effectively before any hardware synthesis must be performed.

STATUS AND OUTLOOK

The design described in this paper has already been implemented in hardware using the description language VHDL. We have performed a successful synthesis of the VHDL design of the EPA, EC, and FD units. We also have verified that the results of FPGA test bench and VHDL simulation match for a number of different input packages. Recently the command and data interface to the RENA chips has been implemented as VHDL design in an FPGA [6]. The next steps will be to verify a correct time tagging of events and to test the functionality with a continuous input data stream. For this purpose the direct connection between the XILINX FPGA board and the RENA board has to be tested and the host computer will only be used as a storage device. Having verified the functionality of the final EPP we will begin to design a "stand alone" FPGA chip without the need for a host computer.

REFERENCES

1. J. Braga, these proceedings
2. R. Jager et al., A&A Suppl. **125**, 192-201 (1997)
3. R. E. Rothschild et al., these proceedings
4. S. Suchy et al., Proc. SPIE **5501**, 312 (2004)
5. T. Schanz, EPP Design Report, Astronomisches Institut der Universität Tübingen (2001)
6. G. Distratis, Diploma thesis, University of Tübingen (2006)

MIRAX Software Aspects

J. Wilms*, S. Schwarzburg[†], R. Remillard**, E. Kendziorra[†], R. Staubert[†] and R. E. Rothschild[‡]

Department of Physics, University of Warwick, Coventry, CV4 7AL, United Kingdom
[†]*IAA Tübingen, Abt. Astronomie, Sand 1, 72076 Tübingen, Germany*
**MIT Kavli Institute, MIT, Cambridge, MA 02139-4307, USA*
[‡]*CASS, UC San Diego, La Jolla, CA 92093-0424, USA*

Abstract. We give a general overview of the design philosophy behind the scientific analysis software for the Hard X-ray Imager on board *MIRAX*. The aim of the software development process is to re-use as much software as possible between the development phase and flight phase of the mission, working on standardized data formats (FITS) and in a standardized, HEASOFT-based software environment. Results from a "proof of concept" study to develop an interactive analysis system for the analysis of realtime (laboratory) data are presented.

Keywords: data acquisition, data analysis
PACS: 01.50.hv,07.05.Hd,07.05.Kf

1. INTRODUCTION

The scientific aims of the *MIRAX* project are to provide astronomers with near real time data about X-ray emitting objects in the Galactic center region. In this contribution, we will give an overview of current developments performed by the authors on the overall design of the scientific analysis software for the Hard X-ray Imager on-board *MIRAX*[1]. We will show that the scientific aims and the needs of scientists and engineers developing the *MIRAX* instruments are in many areas very similar. Because of this similarity, a significant fraction of the required software can be used in both settings, in the laboratory for engineering tests, and during flight. The aim of the software development for *MIRAX* is to make use of this similarity by starting early on in the mission development phase to develop the final software pipelines and to use these pipelines also during the development stage of the mission. This approach of having one code base for both phases of the mission is a significant deviation from the traditional software development approach for X-ray and gamma-ray instruments, where typically two different sets of software are used. It is expected that this new approach results in a more cost effective use of the available resources.

In the remainder of this contribution, we describe these ideas in greater detail. In Sect. 2 we describe the aims of the *MIRAX* software development. We specify the design in Sect. 3, and describe a "proof of concept" study of the design in Sect. 4.

---

[1] For the Soft X-ray Imager, the software analysis system of the *BeppoSAX* Wide Field Camera will be used.

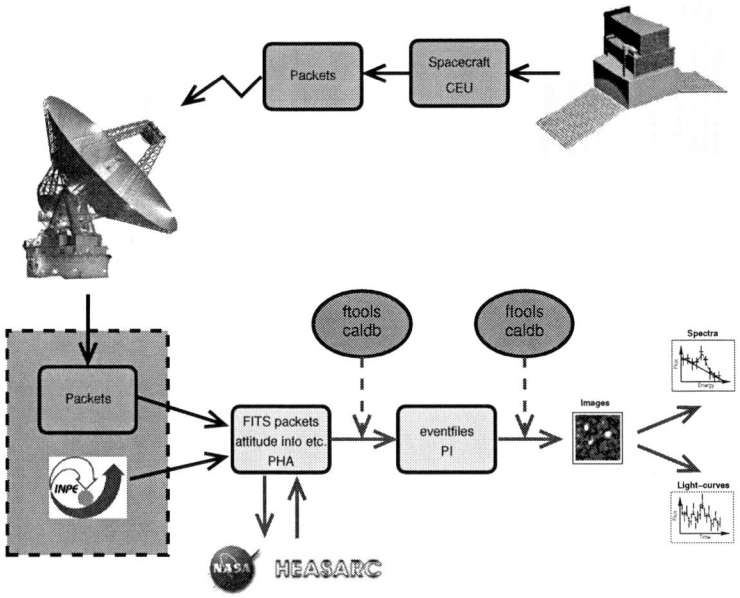

FIGURE 1. Schematic diagram of the data flow for scientific data from the *MIRAX* mission.

2. *MIRAX* SOFTWARE DEVELOPMENT

As mentioned above, the aims of the *MIRAX* software development are to yield a system that can be used for both, the scientific standard analysis and for the analysis of laboratory data. For the scientific standard analysis, the software will have to be able to quickly produce standard data products for dissemination via the Internet (Fig. 1). These results will have to be produced within a short time of the measurements and then be distributed immediately as quick look data to the scientific community. These data products will include images (in standard energy bands), fluxes for individual sources (again, in standard energy bands, and also for different time intervals), spectra of X-ray sources (for standard time intervals), and lightcurves (in standard energy bands and with standard time resolutions). In addition, the software will also allow scientists to analyze the archival *MIRAX* data. This will allow studies where the standard data products are not detailed enough (e.g., creation of spectra using different time intervals than the standard data, pulse phase spectroscopy, etc.).

In addition, the software will have to allow the analysis of laboratory data, especially of calibration measurements and of other measurements aimed towards characterizing the detectors, such as accumulating and analyzing spectra from different regions of the detector. Note that the overall data flow for this process (Fig. 2) is very similar to that of the science analysis (Fig. 1).

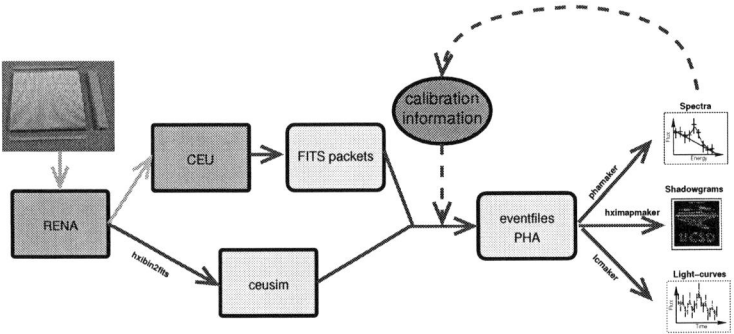

FIGURE 2. Schematic diagram of the data flow for scientific data during the instrument development stage.

3. SOFTWARE DESIGN

The design of the software for the HXI will be based on the heritage from the Rossi X-ray Timing Explorer (*RXTE*). Only standardized data formats and non-proprietary high-level computer languages will be used. These requirements imply

- a software system consisting of individual programs dedicated to individual tasks, instead of a large monolithic software architecture, allowing flexibility in the generation of analysis pipelines due to the capability to use the software from within a scripting language. Note that this implies Linux or other Unix derived operating systems as the major operating system.
- the use of a HEASOFT compatible syntax for the analysis programs, to ensure compatibility with the majority of current X-ray and gamma-ray missions.
- the use of the Flexible Image Transport Standard (FITS; e.g., [1, 2, 3]) as the sole data format for all (intermediate) data products used by the software.

To allow for a cost-effective software development, code re-use will be maximized, by taking the heritage of the mission into account (*RXTE*-ASM, *BeppoSAX*) and by making use of powerful libraries such as HEASARC's `cfitsio` library [4].

4. INTERACTIVE DATA ANALYSIS USING HEASOFT-BASED TOOLS

As illustrated in Figs. 1 and 2, the general data flow during the scientific analysis and during the development of a detector in the lab is rather similar. However, there are major differences in the work flow in both environments. In general, astronomers analyze data in a "batch mode", where scientific data are run through data analysis pipelines and then interpreted. This process might be repeated several times before the data reduction are completed. For example, an astronomer might first extract the full lightcurve for a source, then determine what time intervals are affected by large background count rates, and then rerun the analysis pipeline, excluding this time interval. In contrast, in the laboratory there is far more need that experimenters see the effect of changes to the equipment in real time. For example, if a radioactive source is brought into the field of view of the detector, the count rate displayed on screen should be seen immediately to increase, without first requiring the experimenter to run a separate piece of software. Since typical astronomical data analysis systems do not allow this "near real-time" analysis of experimental data, experimentalists have traditionally resorted to software packages such as Labview or IDL. Unfortunately, this approach has resulted in large amounts of double work, as the knowledge about a detector defined by the Labview or IDL implementation had to be translated for the use of the scientific analysis software.

Based on the requirements for *MIRAX*, we have performed a study whether HEASOFT/FITS based software can in principle also be used for near real time applications. The study is based on our implementation of a data reduction pipeline for an early version of the CdZnTe-detectors that will be used for the HXI and forms the core of the diploma thesis of co-author S. Schwarzburg at the University of Tübingen, Germany [5].

The base for the test is a set of HEASOFT-based programs implementing the whole data reduction process for the HXI. Raw data from the HXI are first repackaged into the FITS format. Then, a second program performs a first polynomial energy correction, determines the detector coordinates for each event, etc., resulting in a first event file. Based on this event file, different dedicated HEASOFT-based tools then generate FITS light curves and PHA-files, which are compatible with the relevant standards [6, 7]. Another program determines detector maps. A final data analysis is then performed using standard astronomical software analysis packages as XSPEC for the spectral analysis.

Because all programs use HEASARC's `cfitsio`-library, it is possible to select detector regions, time intervals, etc. using an extended FITS-file name specification [4]. Even very complicated event selection operations are possible without having to write new code for the HXI-tools. The `cfitsio`-library also allows to copy files over the network (using the HTTP and FTP protocols) and allows storage of FITS-files in shared memory. The latter is of crucial importance for time critical applications.

Using ideas based on MIT's *RXTE*-ASM pipelines, a general pipeline driver for HEASOFT/`cfitsio` based programs was developed. The software allows the general definition of pipelines, based on the properties of their input and output files, using an XML format (see Fig. 3 for a visualization). The pipeline driver executes the relevant programs as soon as data required by these programs are available. Programs can run in parallel. Shared memory is used for maximum speed. The usage of file locking mechanisms and Unix-semaphores ensures maximum speed by allowing the operating system to optimally schedule individual processes and avoids deadlocks within the pipeline. Intraprocess communication between different programs in the pipeline is supported using the XPA messaging standard [8], which is supported, e.g., by visualization programs popular in astronomy, such as `SAOimage` or `ds9`. This allows, for example, the programs to immediately react if the user changes any settings (e.g., time selection, change of thresholds, definition of detector regions for which events are to be accumulated, etc.). For dedicated pipeline software, event loops prevent the cost of multiple startups of a program, standard HEASOFT programs and XSPEC are called via a wrapper program which is responsible for the communication with the pipeline driver.

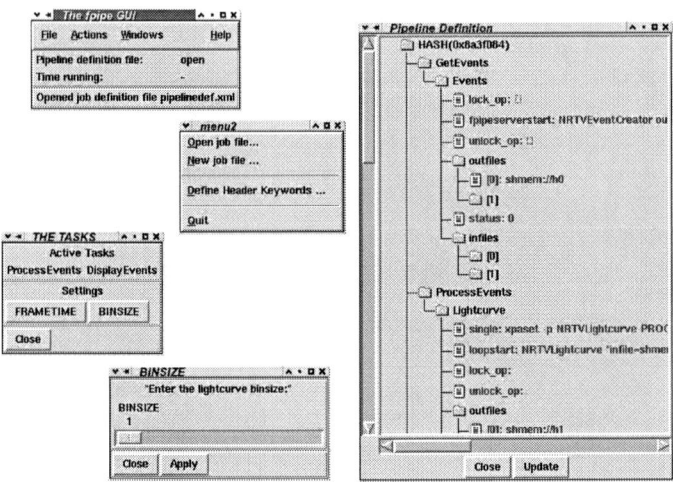

FIGURE 3. Definition and visualization of an analysis pipeline using a graphical user interface.

In tests, pipelines such as the one described here are able to withstand data rates that are typical for a laboratory setup and a real-time analysis of these data is possible. The pipeline therefore constitutes a "proof of concept", showing that software written for the analysis of satellite data can also be used in the laboratory.

For 2006, it is our plan to improve on the existing pipeline driver and software to move the development of the CdZnTe-detectors for the HXI further towards the concepts described in this contribution.

REFERENCES

1. D. C. Wells, E. W. Greisen, and R. H. Harten, *Astron. Astrophys. Suppl. Ser.* **44**, 363 (1981).
2. R. J. Hanisch, A. Farris, E. W. Greisen, W. D. Pence, B. M. Schlesinger, P. J. Teuben, R. W. Thompson, and A. Warnock, *Astron. Astrophys.* **376**, 359–380 (2001).
3. W. D. Cotton, D. Tody, and W. D. Pence, *Astron. Astrophys. Suppl. Ser.* **113**, 159 (1995).
4. W. Pence, "CFITSIO, v2.0: A New Full-Featured Data Interface," in *Astronomical Data Analysis Software and Systems VIII*, edited by D. M. Mehringer, R. L. Plante, and D. A. Roberts, Astron. Soc. Pacific, Conf. Ser. 172, 1999, p. 487.
5. S. Schwarzburg, Eine Software zur Echtzeitanalyse von experimentellen Daten im Flexible Image Transport System, Diploma thesis (2005), available at http://astro.uni-tuebingen.de/publications/diplom.shtml.
6. L. Angelini, W. Pence, and A. Tennant, *Legacy* p. 32 (1993), available online at http://heasarc.gsfc.nasa.gov/docs/heasarc/ofwg/docs/summary/ogip_93_003_summary.html.
7. K. A. Arnaud, I. M. George, and A. F. Tennant, *Legacy* p. 65 (1992), available online at http://heasarc.gsfc.nasa.gov/docs/heasarc/ofwg/docs/summary/ogip_92_007_summary.html.
8. E. Mandel, R. Swick, and D. Tody, "The X Public Access Mechanism," in *Proc. 9th X technical conference*, edited by P. Ferguson, The X Resource 13, O'Reilly and Associates, Inc., Sebastopol, CA, 1995, available at http://hea-www.harvard.edu/RD/xpa/.

On-Board Computing Subsystem for MIRAX: Architectural and Interface Aspects

Valdivino Santiago

Department of Atmospheric and Space Sciences
National Institute for Space Research, São José dos Campos, SP, 12227-010, Brazil

Abstract. This paper presents some proposals of architecture and interfaces among the different types of processing units of MIRAX on-board computing subsystem. MIRAX satellite payload is composed of dedicated computers, two Hard X-Ray cameras and one Soft X-Ray camera (WFC flight spare unit from BeppoSAX satellite). The architectures for the On-Board Computing Subsystem will take into account hardware or software solution of the event pre-processing for CdZnTe detectors. Hardware and software interfaces approaches will be shown and also requirements of on-board memory storage and telemetry will be addressed.

Keywords: interface, computer architecture, on-board computing subsystem, satellite, MIRAX.
PACS: 07.05.-t, 07.87.+v, 89.20.Ff, 89.20.Kk.

INTRODUCTION

The Monitor e Imageador de Raios X (MIRAX) is an X-Ray astronomy satellite mission which the main scientific goals are based in a unique capability of the mission: a hard X-Ray survey of central Galactic plane with Galactic Center continuous monitoring (at least 9 months/year). This mission is currently being developed by the Astrophysics Division (DAS) of the National Institute for Space Research (INPE) in cooperation with research institutes and universities from USA, Germany and The Netherlands [1].

MIRAX payload is composed of two Hard X-Ray cameras (HXI 1 and HXI 2) and one Soft X-Ray camera (SXI), which is WFC flight spare unit from BeppoSAX satellite [2]. A payload computing system is necessary in order to collect and format scientific and kousekeeping data from the CdZnTe detectors of HXIs, to obtain and format data from the SXI, to receive and execute commands from the On-Board Data Handling Computer (OBDH), to trasmit telemetry data to OBDH among other tasks. Without such a computing system, the OBDH would have to interface directly with the X-Ray cameras and this will demand a high computational effort for the platform computer. This paper will present possibilities for the architecture of the computing subsystem and interfaces for MIRAX satellite.

PROPOSALS OF COMPUTING SUBSYSTEM ARCHITECTURES

The proposals of computing subsystem architectures will take into account different aspects like hardware or software solution of the event pre-processing and easiness interfacing of the WFC.

Hardware Event Pre-Processing Option

This option assumes it will be necessary a hardware event pre-processing for CdZnTe detectors of HXIs. This implies the design and implementation of an Event Pre-Processor (EPP) which will perform a fast data processing of detectors raw data to provide usable scientific information [3]. The architecture is shown in Fig. 1.

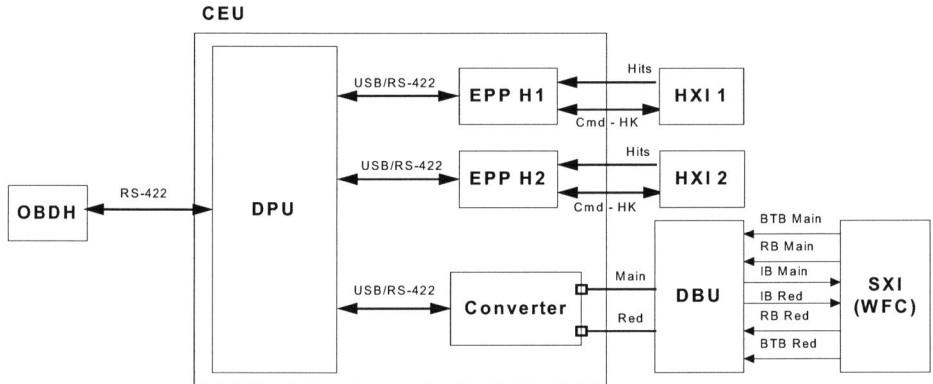

FIGURE 1. Computing subsystem architecture with Hardware Event Pre-Processing option. Legend: Cmd = Command; HK = Housekeeping; Main = Main bus; Red = Redundant bus.

The Central Electronics Unit (CEU) includes the Data Processing Unit (DPU), which is responsible for the communication with the OBDH, obtaining scientific (event packets) and housekeeping data from the EPPs, formatting data, data memory management among other tasks. Also, one EPP will interface directly with one HXI. Each EPP will assemble hits to events, perform error correction of electrical signals, generate scientific and housekeeping data collected from its HXI.

A serial communication line with RS-422 standard will connect the OBDH to the CEU (through DPU). USB or RS-422 serial standards can be used between the DPU and each EPP. USB is a more recent standard that provides a higher data transmission rate than RS-422 one, although RS-422 has a balanced transmission which is better to deal with noise and external interferences.

The OBDH bus in the BeppoSAX satellite was a double redundant one [2]. The WFC connected to this bus via a Digital Bus Unit Interface (DBU). Thus, a Converter is necessary in order to translate the electrical characteristics of the BeppoSAX bus into a USB or RS-422 serial line. Details of the Block Transfer Bus (BTB), the Response Bus (RB) and the Interrogation Bus (IB) can be seen in [2].

The main adavantage of this approach is the use of a dedicated processor (EPP) to interface directly with one HXI. This makes the tasks of the DPU more related to data formatting, management and transmission leaving to the EPPs the work to obtain data from the detectors. The disadvantage is the need for building extra hardware.

Software Event Pre-Processing Option

The main difference between this option and the latter is exactly the absence of EPPs. In the Software Event option, the communication with the HXIs is done by the DPU. Hits goes right to the DPU and the DPU can send commands and obtain housekeeping data from each HXI. The advantage of this approach is exactly not to construct two extra hardwares (EPPs) but the complexity of DPU will increase. Also, it is important to make a performance analysis to see if the DPU system can meet the telemetry (TM) information flows requirements of the application.

WFC Option

In the two options presented so far, nothing was said about the complexity of the Converter (Fig. 1). It can be a difficult task to try to translate the electrical characteristics of the BeppoSAX OBDH bus into the proposed serial standards. One solution is to develop a bus with the same electrical characteristics that the BeppoSAX one so that it was easier to connect the WFC unit. Figure 2 shows this architecture.

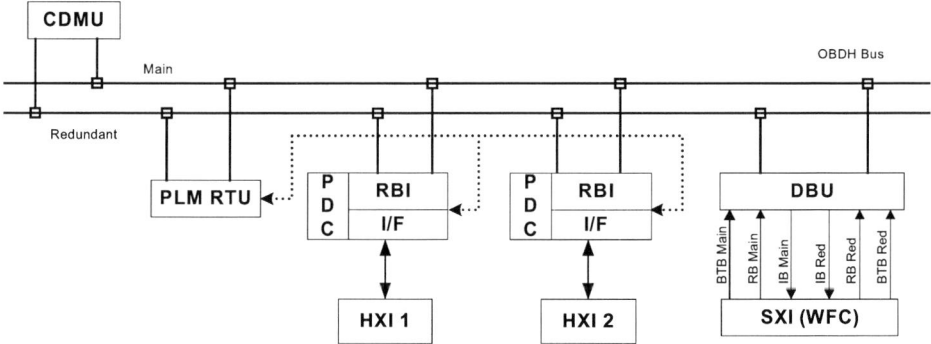

FIGURE 2. Computing subsystem architecture with WFC option.

The Control and Data Management Unit (CDMU) receives and processes telecommand packets, interrogates each Payload Data Handling Computer (PDC) and the WFC for TM packets, formats packets for critical instrument housekeeping parameters acquired by the Payload Module Remote Terminal Unit (PLM RTU), among other tasks. The PDC has the functions of the DPU and the EPP in the Hardware option. The Remote Bus Interface (RBI) makes the connection to the OBDH bus and the I/F is the interface with the HXI.

By using a bus with the same electrical characterisitics of the BeppoSAX one, it is expected to have no much trouble to use the WFC in an architecture like that. But, the complexity of developing the hardware and software of the system in Fig. 2 appears to be much higher than the Hardware and Software Event options.

PROPOSAL OF SOFTWARE INTERFACE

This section will present packet formats as a proposal of software interface for the Hardware and Software Event options presented. Also, a brief discussion about TM requirements will be shown.

The WFC has defined two types of data to be transmitted: Engineering HK data and the BTB TM data [2]. The BTB TM data is a fixed format with 558 16-bit words whch results in 1,116 Bytes per block. Based on this size, a command/response packet between the OBDH and the DPU is proposed in Fig. 3.

SD	OId	NU	Type	CSC	Len	Data	CKS
1	1	4	1	1	2	0 - 1116	2

(a)

SD	OId	Time	Type	RSC	Len	Data	CKS
1	1	4	1	1	2	0 - 1116	2

(b)

FIGURE 3. Proposal of packet formats as software interface between OBDH and DPU. Number below each field represents the size in Bytes. (a) command packet from OBDH to DPU. (b) response packet from DPU to OBDH. NU = Not Used.

The Starter Delimitter (SD) means the beginning of a new command/response packet transmission. The Origin Identificator (OId) field indicates the processing unit that generates the packet. (e.g., OBDH identification for command and DPU, EPP H1, EPP H2 and WFC identification for response packets). Type field refers to different types of commands (e.g., Transmit Scientific Data) and responses (e.g., Scientific Data Packet).

The Command Sequence Control (CSC) allows the DPU to see whether the packets for loading a new program have been received in the correct order. The Response Sequence Control (RSC) field is used by the DPU to indicate to the OBDH how many data packets for a certain type (e.g., scientific) there still exists in the DPU memory.

The Data field has a maximum size of 1,116 Bytes which implies one BTB TM data or 186 event packets, from the EPP/HXI set, per response packet. Actually, each event packet is a 48-bit packet [3]. The Length (Len) field says the amount of Bytes transmitted in the Data field and the 16-bit Checksum (CKS) ends the packets.

It is worth of note the proposed packet formats are established in a communication protocol specification [4] developed in the context of the Qualidade do Software Embarcado em Aplicações (QSSE) project at INPE. The QSSE project is a research project which uses MIRAX as a case study and European Cooperation for Space Standardization (ECSS) standards for software development [5].

About volumes of TM information flows, a preliminary evaluation of the payload demand (HXIs, DPU, WFC) resulted in 150,000 bps total TM rate requirement. This value was achieved considering only the payload demand and assuming one station pass can be missed. Although this a preliminary number and need to be reviewed, it gives a general idea about the TM requirements for MIRAX. By considering a 90-minute orbit, the on-board storage memory requirement is $\approx$ 773 Mbit. Within a 10-minute visibility window, the downlink rate is $\approx$ 1.36 Mbps. At first glance, it is expected that the S-band TM rate of the ground station can be close to 2 Mbps in order to meet the downlink rate requirement satisfactorily.

CONCLUSIONS

Three proposals of on-board computing subsystem architeture for MIRAX were discussed including their interface aspects. The Hardware Event approach allows the DPU tasks to be more related to data management, formatting and transmission to OBDH because the EPPs will perform all the work to communicate with the HXIs. The Software Event option has as main advantage the absence of EPPs with the DPU being responsible for acquiring data from the HXIs cameras. Therefore, it will be a more complex system. Besides that, a performance analysis shall be made in order to see the DPU can meet the TM requirements. The WFC option is the solution where it is proposed the use of a double redundant bus with the same electrical characterisitics of the BeppoSAX one. Although it is expected less trouble to connect the WFC, it seems to be the most demanding option in terms of development. A detailed analysis of these three possibilities shall be made in order to choose the architecture that will demand less time and budget to be accomplished. Proposal of packet formats as software interface between OBDH and DPU were also shown based on a work developed in the QSEE project. From the preliminary study of the TM requirements and downlink rate, it is expected that the S-band TM rate of the ground station can be close to 2 Mbps.

ACKNOWLEDGMENTS

The author would like to thank Financiadora de Estudos e Projetos (FINEP) for supporting the QSEE project.

REFERENCES

1. Braga, J. et al., these proceedings.
2. Alenia Spazio S.p.A., "SAX User's Manual - Volume 11: Wide Field Cameras", Doc. No. SX-MA-A1-019, 1996.
3. Schanz, T., "EPP Design Report", Institute for Astronomy and Astrophysics of the University of Tübingen, 2001.
4. Santiago, V., "Protocolo de Comunicação PDC-OBDH", Doc. No. Q00-PPDOB-v04, QSEE Project, National Institute for Space Research, 2005.
5. European Cooperation for Space Standardization, "Software – Part 1: Principles and Requirements" in *ECSS Space Engineering*, Doc. No. ECSS-E-40 Part 1B, 2003.

Payload Software Validation and Integration

Maria de Fátima Mattiello-Francisco

Brazilian National Institute for Space Research INPE
Av. dos Astronautas 1758, CEP 12 201970
São José dos Campos, São Paulo
Brazil

Abstract. An *operational profile* describes how users employ a system. We apply this technique aiming at an effort reduction in both time and cost on the process of validation and integration of the X-ray imager software embedded in the *MIRAX* satellite mission payload. Once the scientific goals of the payload instrument are well defined by the mission principal investigators, the operation profiles help engineers and software designers to identify the payload development resources and testing workbench to guarantee that the instrument and its operation meet the scientific requirements. One benefit is to reuse some test facilities designed to support the incremental development of the payload instrument functionalities on further satellite mission operations. Another benefit is the system growth in reliability, since testing driven by operational profiles helps to anticipate failure identifications.

Keywords: software engineering, testing facilities, operational profile, satellite payload software
PACS: 01.50.hv, 01.50.Kw, 07.05.Bx, 07.05.Hd

SATELLITE MISSION LIFECYCLE

In space missions, a software product is part of a network of systems. In particular, satellite payload embedded software is a subsystem of the satellite mission in which the development and product qualification phases involve a complex process for validation and integration.

Over the years a number of models have been devised to describe the various phases of a project. The models identify the components of the project and may indicate interdependencies or interrelationships. Although each of the models has its own characteristics and advantages, they often take into account the lifecycle of the product.

The model used by the ECSS – European Committee Space Standardization - for space mission development is based on a typical mission lifecycle. It comprises 6 phases: 0+A- Mission Analysis/Needs Identification and Feasibility, B- Preliminary Definition (Project and Product), C- Detailed Definition (Product), D- Production/ Ground Qualification Testing, E- Utilization, F- Disposal [1], along which 7 main activities are carried on: mission conception, requirements definition, technical specification, verification and qualification, production, operation and disposal. Fig. 1 presents ECSS the basic mission lifecycle and baselines created by ESA and the space industry in Europe to involve the industry in the execution of such activities as space program suppliers. It highlights the relationship of those space mission activities with the activities of a typical software lifecycle processes, described in the standard ECSS-E-40 Part 1B [2], which takes into account the existing ISO 9000 family of documents and the ISO/IEC 12207 standards [3]. The scratched bars represent the period in which each space mission activity is carried out from the system point of view. The filled bars correspond to the implementation of such activities related to the space software subsystem development.

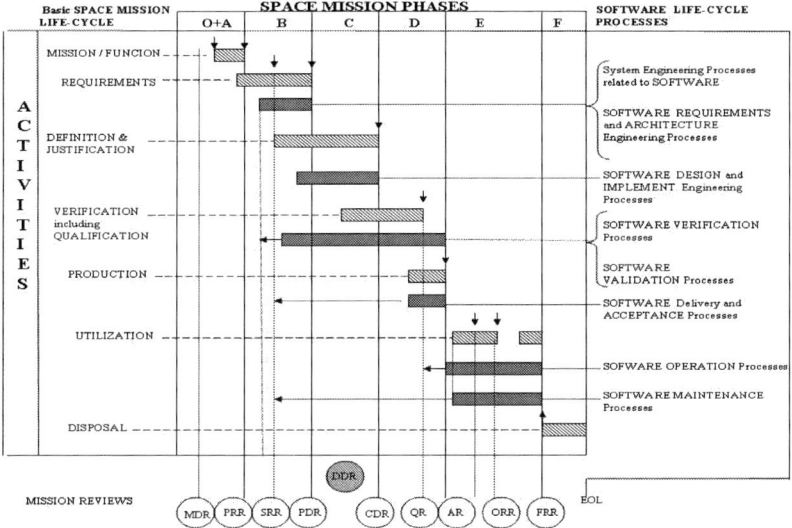

FIGURE 1. ECSS typical project lifecycle. **Bottom**: shows the typical sequence of reviews followed in the space mission project: MDR- Mission Definition Review, PRR- Preliminary Requirements Review, SRR- System Requirement Review, PDR- Preliminary Design Review, CDR- Critical Design Review, QR- Qualification Review, AR- Acceptance Review, ORR- Operation Requirements Review, FRR- Flight Readiness Review. Since software project is usually considered to be a subsystem in the hierarchy of space system development, it shall be synchronized with those milestones. Moreover, depending on the complexity of the software, the reviews in gray may be replicated at the subsystem development level. DDR- Detailed Design Review is particularly recommended for software projects.

In the case of satellite missions, another important issue along the mission development phases is the production of the 3 satellite models: engineering model, qualification model and flight model. All of them demand a quite complex set of test facilities to be evaluated.

Since the engineering area at INPE responsible for the development of the satellite platform follows ECSS recommendations in the scientific satellite development, our purpose is to apply the ECSS in the MIRAX payload software. The tailoring of ECSS software process requirements for MIRAX payload software got started last year in the QSEE (Quality improvement on space application embedded software) project. As a case study, a downsized of MIRAX software payload has been developed [7].

INTEGRATION AND VALIDATION

System integration involves assembling the complete system from its component modules and performing initial testing to verify its functionality before progressing to full system testing and validation. A range of methods is used to achieve integration, although progressive integrations are the more traditional approach [4]. Software payload integration in satellite missions developed by INPE follows such evolutionary process that requires particular testing resources for each integration stage.

Since validation is the process of determining that a system or subsystem meets its requirements, the complete validation of satellite payload embedded software occurs only when the payload instrument can be remotely controlled from the mission ground segment.

Usually, in a satellite mission's development at INPE, 3 integrated testing stages are performed: at the instrument level, at the satellite subsystem level and at the system level. The diagrams (a), (b) and (c) in Fig. 2 represent, respectively, the hardware/ software modules and the testing workbench elements required for the payload software integration and validation at each stage. System integration testing investigates the characteristics of a collection of modules and is generally aimed at establishing their correct interaction. It comprises a search for faults that cause inter-component failures, focusing on interoperability.

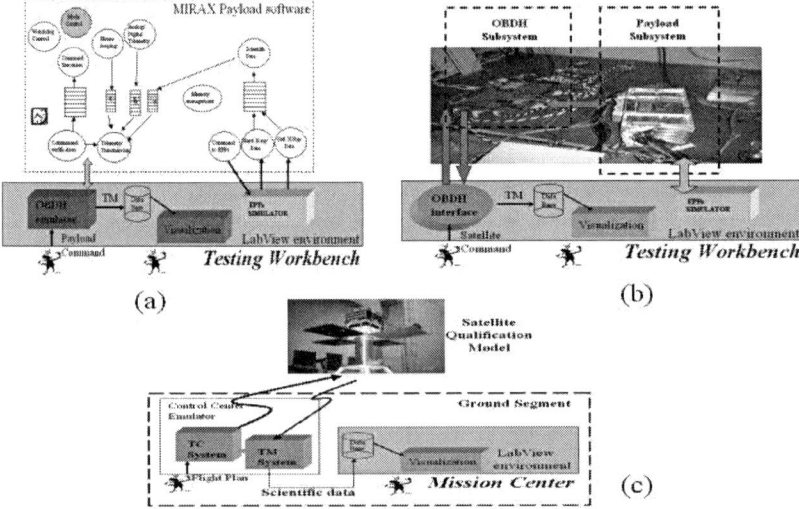

FIGURE 2. **(a)**: A simple representation of the MIRAX payload software components and testing workbench elements required for integration and validation at instrument level; **(b)**: testing workbench for satellite payload and OBDH computer integration and validation at satellite platform subsystem level; **(c)**: a simple view of the testing facilities for software integration and validation at system level using the satellite qualification model.

The instrument level deals with the integration of all payload components (hardware and software) in order to evaluate the instrument performance and conformance with the scientific goals of the mission. For this purpose, a specific testing workbench, composed by an on-board data handling emulator, sensor simulators (events acquisition and pre-processing EPP) and a testing system to control the test execution and data visualization/ analysis, shall be provided.

At the subsystem level, the integration testing focus on the payload communication with the satellite platform, more specifically with the On Board Data Handling computer – OBDH, in order to validate payload behavior when the actual OBDH computer and other satellite subsystems are in the loop. For this purpose, obviously the OBDH is replaced by the actual hardware; however, the payload sensor simulators can be reused and the testing system can be tailored to control the test execution. In this way, the testing workbench to validate the payload behavior shall provide a communication interface with the actual OBDH to send commands to the payload and receive its telemetry so that the data visualization/ analysis tool can be reused.

The purpose of system level integration is to validate the integrity of ground-board satellite interaction as well as the procedural contents of the flight operation plan. The payload is already integrated in the satellite and the communication with it is done only through ground station and satellite telecommunication subsystems. At this stage the testing workbench shall include the ground station testing valise which includes control center facilities for sending commands to the satellite payload and receiving scientific data. Again the data visualization/ analysis tool can be reused.

OPERATIONAL PROFILES

The operational profile is an external user-oriented test model, which specifies the intended usage of the system in terms of events and their invocation probabilities [5, 6]. By providing information about how users will employ the software product, operational profile supports software engineering team on the resources definition for development and testing.

Developing an operational profile to guide testing involves as many as five steps: find the customer profile, establish the user profile, define the system-mode profile, determine the functional profile and determine the operational profile itself. Some steps may not be necessary in a particular application. The first four steps break down the system use progressively into more detail. In the last one, functions evolve into operations as the system is implemented [5].

In the case of satellite payloads, one can break down the embedded software system into user groups responsible for the instruments which compose the payload. A single user group may invoke several system modes. A system mode is a set of functions or operations that an engineering team groups for convenience in analyzing execution behavior. A system can switch among modes so that only one is in effect at a time, or it can allow several modes to exist simultaneously, sharing the same resource. A system mode profile is the set of system modes and their associated occurrence probabilities. In turn, each system mode has several functions, for instance, the command verification and telemetry data transmission functions in nominal mode. Functional profile information such as number of functions, functions list, environmental variables and each function's occurrence probability provides a quantitative picture of the relative use of different functions.

The functional profile is a user-oriented profile of functions, not the operations that actually implement them. The goal is to test operations. An *operation* is a major system logical task (related to a functional requirement or feature of a product, not a subtask in the design) of short duration, which returns control to the system when complete and whose processing is substantially different from other operation. Functions evolve into operations as the operational architecture of the system is developed. The *operational architecture* is the way the user will employ operations to accomplish functions. There is often some but rarely a complete correlation between the operational architecture and the system architecture, thus the mapping from functions to operations is not necessarily straightforward. Generally there are more operations than functions, and operations tend to be more refined. For example: the telemetry data transmission function mentioned above is implemented by at least two operations, packing telemetry data and sending those packets to the OBDH computer. An operation is usually more differentiated than a function, in that it represents a particular task with certain specific input-variable values or value ranges.

The *profile* is simply a set of disjoint (only one can occur at a time) alternatives with the probability that each will occur. If A occurs 60 percent of the time and B 40 percent, for example, the operational profile is A, 0.6 and B, 0.4. Therefore, the *operation profile* is simply the set of operations and their probabilities of occurrence.

In order to develop an operational profile properly, system engineers and software designers must communicate with the users in a disciplined way that tends to bring out all the operations they need and their probabilities of occurrence. Also, system testers shall participate because they can ensure that testing needs are met.

Testing driven by an operational profile is very efficient because it anticipates operation failures (and hence the faults causing them) early in the development process. It rapidly increases reliability (reduces failure intensity) per unit of execution time because the failures that occur most frequently are caused by the faulty operations used most frequently.

Although operational modes and operational profiles can be further refined with time, initial versions are essential during the requirements phases in order to use them for performance analysis. To allocate testing effort, select tests, and determine the order in which tests should be run, a version of the operational profile that is close to the final must be available when you start to plan the tests.

In order to guide software validation and integration testing in the MIRAX satellite mission, operational profile techniques have been used in the QSEE project case study.

REFERENCES

1. Space Project Management – Project Phasing and Planning, ECSS-M-30A, April 1996.
2. Space Engineering – Software – part 1: Principles and requirements, ECSS-M-40 Part 1B, November 2003.
3. ISO/IEC 12207 Information Technology - Software Lifecycle Processes.
4. Neil Storey, *Safety-critical Computer Systems* – First Edition, 1996.
5. Musa, J. D., "Operational Profiles in Software Reliability Engineering", *IEEE Software*, pp. 14–32, March 1993.
6. Lyu, M. R., (ed.), *Handbook of Software Reliability Engineering*, McGraw-Hill, 1995.
7. http://www.cea.inpe.br/~qsee, QSEE- Qualidade do Software Embarcado e aplicações espaciais, funded by FINEP, 2005.

Workshop

"The Transient Milky Way: A Perspective for MIRAX"

Instituto Nacional de Pesquisas Espaciais – INPE – Brazil

December 7 – 9, 2005

FINAL PROGRAMME

December 7 – Wednesday

08:00 REGISTRATION
08:40 Gilberto Câmara (INPE Director), João Braga (CEA Coordinator) – *Opening*

Session 1: Brazilian space program and MIRAX science overview
Chair: J. Braga

09:00 H. Carvalho – *Brazilian space program* (30 min)
09:30 M. A. Chamon – *INPE's scientific satellite program* (20 min)
09:50 R. Remillard – *MIRAX science overview* (40 min)

10:30 COFFEE BREAK

Session 2: X-ray bursts, Low-Mass X-Ray Binaries, Galactic Center field
Chair: J. Heise

11:00 J. in 't Zand – *The BeppoSAX WFC observation program on the GC field* (35 min)
11:40 A. Cumming – *What can we learn from long-term monitoring of X-ray bursters?* (35 min)

12:20 LUNCH

14:00 C. Markwardt – *RXTE and Swift observations of GC transients* (35 min)
14:40 D. Galloway – *Observations of transient pulsars* (25 min)
15:10 J. Kennea – *Observations of transients by Swift* (25 min)

15:40 COFFEE BREAK

Session 3: Optical/IR/radio observations of X-ray binaries
Chair: F. Jablonski

16:00 E. Gallo – *Jets in Galactic X-ray transients* (25 min)
16:30 R. Hynes – *Multiwavelength variability in transient black hole binaries* (25 min)
17:00 F. D'Amico – *Infrared observations of IGR J16358−4726* (15 min)

17:20 RECEPTION

December 8 – Thursday

Session 3 (cont.): Optical/IR/radio observations of X-ray binaries

09:00   B. Becker – *MAGPIS: The Multi-Array Galactic Plane Imaging Survey* (25 min)
09:30   C. Bailyn – *Optical and IR monitoring of X-ray binaries* (35 min)

10:10   COFFEE BREAK

Session 4: Galactic Plane Monitoring
CHAIR: J. in 't Zand

10:30   W. Paciesas – *BATSE monitoring of the GC region: a nine year history of hard X-ray transient activity* (25 min)
11:00   E. Kuulkers – *INTEGRAL GC monitoring program* (25 min)
11:30   R. Walter – *Results of the INTEGRAL survey of the Galaxy* (25 min)
12:00   J. Tomsick – *INTEGRAL obscured sources* (25 min)

12:30   LUNCH

Session 5: High-Mass X-Ray Binaries, neutron stars and obscured sources
CHAIR: R. Rothschild

14:00   A. Santangelo – *HMXB pulsars* (25 min)
14:30   J. Wilms – *Monitoring neutron stars with INTEGRAL* (25 min)
15:00   R. Staubert – *New results on Her X-1 (RXTE/INTEGRAL)* (25 min)

15:30   COFFEE BREAK

Session 6: Black Holes, Gamma Ray Bursts and fast transients
CHAIR: R. Remillard

16:00   J. Heise – *Fast X-ray transients* (35 min)
16:40   C. Markwardt – *Swift observations of Gamma Ray Bursts* (25 min)
17:10   R. Opher – *How do black holes provide the emitted energy of GRBs, microquasars, quasars and AGNs?* (20 min)

19:00   HAPPY HOUR

Workshop "The Transient Milky Way: A Perspective for MIRAX"

December 9 – Friday

Session 7: MIRAX instruments and software
    Chair: R. Staubert
08:30 J. Braga – *MIRAX mission overview and status* (25 min)
09:00 R. Rothschild – *CZT detectors and HXI development at CASS/UCSD* (25 min)
09:30 J. Mejia – *HXI imaging simulations and sensitivity* (25 min)
10:00 W. Mels – *A wide field X-ray camera for MIRAX* (25 min)

10:30   coffee break

10:45 E. Kendziorra – *Onbord Intelligence: CEU Hard- and Software* (25 min)
11:15 J. Wilms – *Concept for laboratory and science software* (20 min)
11:40 F. Mattiello – *Payload software validation and integration* (20 min)
12:05 V. Santiago – *Payload computing subsystem for MIRAX* (20 min)

12:30   lunch

Session 8: MIRAX spacecraft
    Chair: E. Kendziorra
14:00 A. Silva – *MIRAX preliminary mission analysis* (20 min)
14:25 M. A. Chamon – *MIRAX spacecraft architecture* (20 min)
14:50 Discussion about MIRAX spacecraft, payload and interfaces

15:30   coffee break

15:45 MIRAX team meeting – workshop wrap-up, definition of action items
17:00 *Closing*

Workshop "The Transient Mily Way: A Perspective for MIRAX"

LIST OF PARTICIPANTS

Name	Institution	Country	e-mail
Odylio D. AGUIAR	INPE	Brazil	odylio@das.inpe.br
Mario C. P. ALMEIDA	INPE	Brazil	mcelso@das.inpe.br
José C. N. de ARAUJO	INPE	Brazil	jcarlos@das.inpe.br
Charles BAILYN	Yale University	USA	bailyn@astro.yale.edu
Beatriz BARBUY	IAG/USP	Brazil	barbuy@astro.iag.usp.br
Robert BECKER	UC-Davis/LLNL	USA	bob@igpp.ucllnl.org
Armando BERNUI	INPE	Brazil	bernui@das.inpe.br
João BRAGA	INPE	Brazil	braga@das.inpe.br
Cláudio BRANDÃO	INPE	Brazil	claudio@das.inpe.br
Hugo CAPELATO	INPE	Brazil	hugo@das.inpe.br
Himilcon CARVALHO	AEB/MCT	Brazil	himilcon@aeb.gov.br
Reinaldo de CARVALHO	INPE	Brazil	reinaldo@das.inpe.br
Marco A. CHAMON	INPE	Brazil	chamon@dss.inpe.br
Deonisio CIESLINSKI	INPE	Brazil	deo@das.inpe.br
Andrew CUMMING	McGill U.	Canada	cumming@physics.mcgill.ca
Flavio D'AMICO	INPE	Brazil	damico@das.inpe.br
Marcos DIAZ	IAG/USP	Brazil	marcos@astro.iag.usp.br
Ijar M. da FONSECA	INPE	Brazil	ijar@dem.inpe.br
Raphael FONSECA	INPE	Brazil	raphael@iss.inpe.br
Elena GALLO	UC St. Barbara	USA	egallo@science.uva.nl
Duncan K. GALLOWAY	U. of Melbourne	Australia	D.Galloway@physics.unimelb.edu.au
John HEISE	SRON	The Netherlands	j.heise@sron.nl
Gabriel R. HICKEL	UNIVAP	Brazil	hickel@univap.br
Jorge E. HORVATH	IAG/USP	Brazil	foton@astro.iag.usp.br
Robert I. HYNES	LSU	USA	rih@phys.lsu.edu
Thais E. P. IDIART	IAG/USP	Brazil	thais@astro.iag.usp.br
Francisco JABLONSKI	INPE	Brazil	chico@das.inpe.br
Udaya B. JAYANTHI	INPE	Brazil	jayanthi@das.inpe.br
Eckhard KENDZIORRA	IAAT	Germany	kendziorra@astro.uni-tuebingen.de
Jamie A. KENNEA	Penn Sate U.	USA	kennea@astro.psu.edu
Erik KUULKERS	ESAC/ESA	Spain	Erik.Kuulkers@esa.int
Craig B. MARKWARDT	NASA/GSFC	USA	craigm@milkyway.gsfc.nasa.gov
Fátima MATTIELLO	INPE	Brazil	fatima@iss.inpe.br
Jorge MEJÍA	INPE	Brazil	mejia@das.inpe.br
Manoel de MORAES Jr.	INPE	Brazil	mcvmjr@astro.iag.usp.br
Wim W. A. MELS	SRON	The Netherlands	w.mels@sron.nl

LIST OF PARTICIPANTS (cont.)

Name	Institution	Country	e-mail
Ana F. NASCIMENTO	UNICAMP	Brazil	anafnascimento@yahoo.com.br
Reuven OPHER	IAG/USP	Brazil	opher@astro.iag.usp.br
William S. PACIESAS	UAH/NSSTC	USA	bill.paciesas@msfc.nasa.gov
Adriana M. PIRES	IAG/USP	Brazil	apires@astro.iag.usp.br
Ronald A. REMILLARD	MKI/MIT	USA	rr@space.mit.edu
Fabiola M. A. RIBEIRO	IAG/USP	Brazil	fabiola@astro.iag.usp.br
Milton S. R. RIBEIRO	INPE	Brazil	milton@dmf.inpe.br
Richard ROTHSCHILD	CASS/UCSD	USA	rrothschild@ucsd.edu
Jayashree ROY	Gauhati U.	India	jayashree_ntl@yahoo.com
Andrea SANTANGELO	IAAT	Germany	andrea.santangelo@uni-tuebingen.de
Denilson P. SANTOS	INPE	Brazil	denilson@dem.inpe.br
Valdivino SANTIAGO	INPE	Brazil	valdivino@das.inpe.br
Adenilson R. SILVA	INPE	Brazil	Adenilson.Silva@dss.inpe.br
Adriana V. R. SILVA	CRAAM	Brazil	asilva@craam.mackenzie.br
André B. SILVA	INPE	Brazil	andrebcs@dea.inpe.br
Clóvis SOLANO Pereira	INPE	Brazil	clovis@lit.inpe.br
José P. M. de SOUZA	GISPLAN	Brazil	patrocinio@gisplan.com.br
Rüdiger STAUBERT	IAAT	Germany	staubert@astro.uni-tuebingen.de
Cesar STRAUSS	INPE	Brazil	cstrauss@cea.inpe.br
Kenny C.-TALAVERA	INPE	Brazil	kenny@das.inpe.br
Kamal K. TANTI	Gauhati U.	India	kamal_tanti@yahoo.co.in
Francisco C. TERCEIRO	INPE	Brazil	amorim@dem.inpe.br
John A. TOMSICK	CASS/UCSD	USA	jtomsick@ucsd.edu
Ricardo VARELA Correa	INPE	Brazil	ricardo@cea.inpe.br
Haroldo C. VELHO	INPE	Brazil	haroldo@lac.inpe.br
José VILAS-BOAS	INPE	Brazil	jboas@das.inpe.br
Cláudia VILEGA R.	INPE	Brazil	claudiavr@das.inpe.br
Valeri VLASSOV	INPE	Brazil	vlassov@dem.inpe.br
Roland WALTER	INTEGRAL SDC	Switzerland	Roland.Walter@obs.unige.ch
Jörn WILMS	U. Warwick	United Kingdom	j.wilms@warwick.ac.uk
Carlos A. WUENSCHE	INPE	Brazil	alex@das.inpe.br
Wilson YAMAGUTI	INPE	Brazil	yamaguti@dss.inpe.br
Jean in 't ZAND	SRON	The Netherlands	J.J.M.in.t.Zand@sron.nl

AUTHOR INDEX

B

Becker, R. H., 102
Braga, J., 3, 112
Brandt, S., 30

C

Chamon, M. A., 18
Chenevez, J., 30
Cieslinski, D., 97
Courvoisier, T. J.-L., 30
Cumming, A., 55

D

D'Amico, F., 97
de Carvalho, T. R., 18
de Fátima Mattiello-Francisco, M., 131
Distratis, G., 117

E

Ebisawa, K., 30

F

Fender, R., 83

G

Gallo, E., 83
Galloway, D. K., 50

H

Heise, J., 8
Helfand, D. J., 102
Hickel, G., 97
Hynes, R. I., 88

I

in 't Zand, J., 8

J

Jablonski, F., 93, 97

K

Kendziorra, E., 117, 121
Kennea, J. A., 71
Klochkov, D., 65
Kretschmar, P., 30
Kuulkers, E., 30

M

Markwardt, C. B., 30, 45
Matteson, J. L., 107
Mejia, J., 112
Mels, W., 8
Mowlavi, N., 30

O

Oosterbroek, T., 30
Opher, R., 76
Orr, A., 30

P

Paizis, A., 30
Pelling, M. R., 107
Postnov, K., 65

R

Ramos, L. A., 93
Remillard, R., 121
Rodrigues, C. V., 97
Rothschild, R. E., 107, 121

S

Sanchez-Fernandez, C., 30
Santangelo, A., 60
Santiago, V., 126
Schandl, S., 65
Schanz, T., 117
Schwarzburg, S., 121
Shakura, N., 65
Shaw, S. E., 30
Staubert, R., 65, 121
Suchy, S., 107, 117

T

Tomsick, J. A., 25, 107

W

Walter, R., 35
White, R. L., 102
Wijnands, R., 30
Wilms, J., 40, 65, 121